THEORY OF LINEAR SYSTEMS

Staff of Research and Education Association

Research and Education Association
505 Eighth Avenue
New York, N. Y. 10018

THEORY OF LINEAR SYSTEMS

Printed in the United States of America

Library of Congress Catalog Card Number 82-80746

International Standard Book Number 0-87891-539-7

PREFACE

This book is an excellent reference source for those who possess mathematical skills and who wish to apply those skills to important problems in science and engineering.

The theory of linear systems is widely used in solving problems occurring in fields such as automatic control, communications, optics, statistical systems, and biomedical applications. These fields are currently being widely explored, are growing at a rapid pace, and are regarded as the technologies of the future. There is considerable demand for persons who are able to apply their mathematical skills for solving problems in these areas.

Chapter 1 deals with the fundamental concepts upon which the definition of a system can be established. Such notions as ordered pairs, oriented abstract objects, etc., serve as the basics of the definition. The system response is defined in terms of an input-output-state relationship which must satisfy certain consistency conditions.

The system definition of Chapter 1 is built upon in Chapters 2, 3 and 4. In Chapter 2 the conditions for consistency are recast in the form of demonstrating the separation property inherent in the input-output-state relationship. In Chapter 3, the state space representation is formalized for linear differential systems, and Chapter 4 highlights the canonical form of the state space representation, i. e.

$$\dot{x} = Ax + Bu$$

$$y = Cx + Du$$

In Chapter 5 solutions to the canonical equations are presented for a variety of systems. A natural fall-out of the state space formulation are the theoretical concepts of controllability, observability and stability, which are addressed in Chapter 6.

In the last two chapters the reader is reminded that the linear theory developed thus far has limited application to real world problems, such as treating large complex systems. The objective of Chapter 7 is to introduce the idea of an imprecisely defined (or probabilistic) system. The methods of extracting signals from noise, both of which are treated as random processes, are discussed as a means to augment the linear theory. Finally, Chapter 8, which deals with quantized systems, ends this volume by highlighting the perturbation method for treating state transition probabilities. It exemplifies the high degree of difficulty encountered in analyzing statistical systems.

The book is intended as a useful basic reference, and for this reason detailed proofs of theorems are for the most part omitted.

The information in this book was originated and sponsored by the U. S. Naval Electronic Systems Command and edited by R. Fratila.

CONTENTS

1
Fundamental Concepts and Definitions

1.1 INTRODUCTION

Linear system theory is intended as a discipline to provide a unified conceptual framework for system analysis. In establishing this framework we will introduce such notions as abstract objects, their measurable attributes, and the mathematical relations between attributes. These concepts will serve as the primitives of the theory, although they defy precise definition in unequivocal terms. Our goal will be to identify a small part of the physical world we intuitively understand and attempt to evolve a quantitative basis for analysis. The proof of this quantitative understanding will be manifest in our ability to better describe the behavior of linear differential systems.

We think of a system (in vague terms) as being a collection of things or objects which somehow are united through interaction or interdependence. More precisely, we define[1] a system to be a partially interconnected set of abstract

1. After Zadeh and Desoer, Linear System Theory, McGraw-Hill, New York, 1963, pp. 1-65.

objects, which are called the system components. These components may be oriented or nonoriented; they may be finite or infinite in number; and each of them may be associated with a finite or infinite number of terminal variables. It is not intended or expected at this time that the above definition be fully comprehended by the reader. The terms used to define the system are themselves undefined. Rather, the intent here is to set the stage for the discussion to follow. The remainder of the chapter will be devoted to clarifying and concretizing the concept of a system, and some of its ramifications.

1.2 TIME FUNCTIONS

As a foundation for the sequel our initial goal is to establish the terms of reference, particularly those addressing time dependent variables. Consider an object (as yet undefined) which we will label $\mathcal{A}$, and let u be a measurable attribute of $\mathcal{A}$. u can be real or complex. Let it be understood that u is a time function where time ranges from $-\infty$ to ∞. Let T be a specific subset of time and t an element within T. (Typically, T may be the semi-infinite interval $(t_0,\infty]$, where t_0 is a particular value of t; or T may be a finite interval $[t_0,t_1]$, etc.) The time function u defined on the subset T will be denoted simply as u or as $u[t_0,t_1]$, where $[t_0,t_1]$ is the time segment T. On the other hand $u(t)$ will denote the value of u at time t. Thus, u is meant as the entire set of pairs $\{(t,u(t))\}$ for each t in T. In the specific time segment $[t_0,t_1]$ the function $u[t_0,t_1]$ is the totality of pairs $\{(t,u(t))\}$ with $t_0 \leqslant t \leqslant t_1$. Figure 1.1 further illustrates what is meant by $u[t_0,t_1]$ and $u(t)$ for real-valued variables.

In general the discussions to follow will involve the set or class of time functions $\{u\}$ more so than a single time function u. In the case u may be regarded as a real or complex variable ranging over $\{u\}$. The range of u will be denoted as $R[u]$; it is the set of time functions to which u belongs. When u

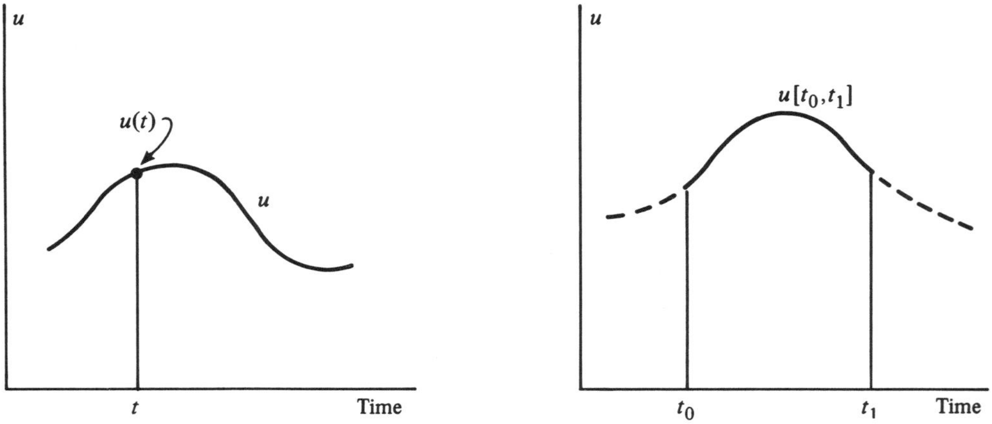

Figure 1.1 Notation for time functions.

varies over $R[u]$, the values of u for fixed t vary over a set which is the range of the variable $u(t)$. The range of $u(t)$ will be denoted by $R[u(t)]$. In general $R[u(t)]$ is independent of t. Figure 1.2 further illustrates $R[u(t)]$ and $R[u]$ for real-valued variables. The distinction between $R[u(t)]$ and $R[u]$ is that the former includes the set of real numbers over $(-\infty,\infty)$, whereas the latter includes a set of time functions.

The discussion for scalar time functions can readily be extended to include vector (and matrix) functions. In bold face $\mathbf{u}(t)$ denotes an n-vector. The range R^n of $\mathbf{u}(t)$ is the space of n-tuples of real numbers whose values are assumed at time t. (The range C^n will be used to denote the set of ordered n-tuples where the elements of $\mathbf{u}(t)$ are complex numbers.) We have

$$\mathbf{u}(t) = (u_1(t),u_2(t),\ldots,u_n(t)) \tag{1.1}$$

where u_i $(i = 1,\ldots,n)$ are real or complex numbers and the parameter t identifies the values of the u_i at time t.

The vector function $\mathbf{u}$ is the totality of pairs $\{(t,\mathbf{u}(t))\}$ for each t in T. For the specified time interval $[t_0,t_1]$ the vector $\mathbf{u}[t_0,t_1]$ consists of the pairs $\{(t,\mathbf{u}(t))\}$, where $t_0 \leqslant t \leqslant t_1$. In the sequel we will generally denote the vector function $\mathbf{u}[t_0,t_1]$ simply as $\mathbf{u}$, i.e.,

$$\mathbf{u} = \mathbf{u}[t_0,t_1] = \begin{Bmatrix} (u_1(t_0),u_2(t_0),\ldots,u_n(t_0)) \\ (\ldots\ldots\ldots\ldots\ldots\ldots\ldots\ldots) \\ (u_1(t_1),u_2(t_1),\ldots,u_n(t_1)) \end{Bmatrix} \tag{1.2}$$

where t_0 and t_1 are the parameters associated with the time segment T.

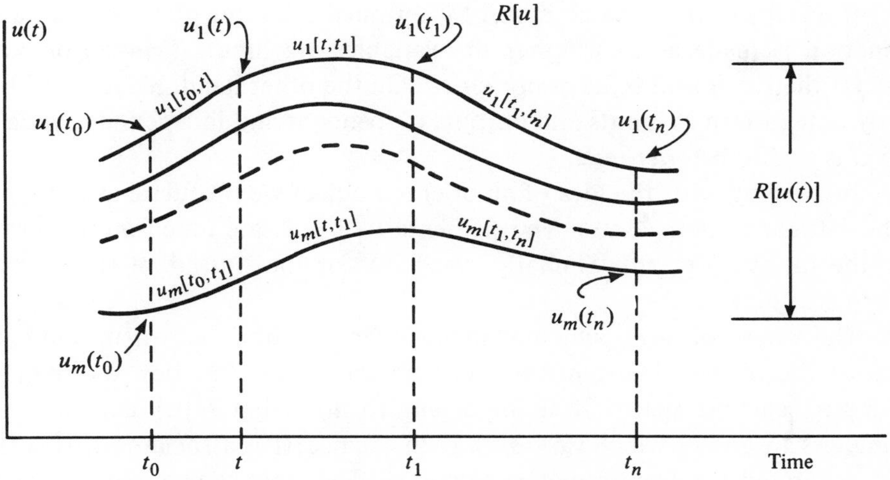

Figure 1.2 Notation for range of time functions.

Typically, the vector **u** is a vector-valued time function which, (a) may contain a finite number of delta functions of various orders over a finite interval, and (b) is piecewise continuous and has piecewise continuous derivatives of finite orders on every finite interval over which it has no delta functions. Thus, it will be assumed that $R[\mathbf{u}(t)]$ is R^n and $R[\mathbf{u}]$ is the space of all vector-valued time functions which are finitely differentiable over the time segments T.

1.3 OBJECTS, ATTRIBUTES AND TERMINAL RELATIONS

An object can be labeled as either physical $\mathcal{P}$ or abstract $\mathcal{Q}$. (Within these categories we can further classify the objects as either oriented or non-oriented.) By a physical object it is usually meant a physical device which is associated with a set of attributes $u_1, u_2, u_3, \ldots,$ where the relations between these attributes necessarily characterize the object. In effect, an object is a set of variables with a defined set of relations between them. Specifically, these variables are called the terminal variables of object $\mathcal{Q}$; the relations between them are the terminal relations. The characterization of $\mathcal{Q}$ by its terminal relations can be written symbolically as

$$\begin{aligned} \mathcal{Q}^1(u_1,u_2,\ldots,u_n) &= 0 \\ \mathcal{Q}^2(u_1,u_2,\ldots,u_n) &= 0 \\ \cdots\cdots\cdots\cdots\cdots\cdots \\ \mathcal{Q}^m(u_1,u_2,\ldots,u_n) &= 0 \end{aligned} \tag{1.3}$$

where each $\mathcal{Q}^j$, $j = 1,2,\ldots,m$, represents a relation between the variables u_i, $i = 1,2,\ldots,n$.

If an object $\mathcal{Q}$ is characterized by terminal relations of form (1.3) and no distinction is made as to whether the variables are inputs (causes) or outputs (effects) then $\mathcal{Q}$ is said to be *nonoriented*. On the other hand, if the variables are clearly categorized as inputs and outputs (dependent and independent variables), then $\mathcal{Q}$ is said to be *oriented*.

In keeping with the idea of an oriented object we postulate as the input to the k-tuple $\mathbf{u} = (u_1,\ldots u_k)$. The elements $u_1,u_2,\ldots$ are time functions varying over the interval (t_0,t_1). Similarly, we have for the output of $\mathcal{Q}$ the m-tuple $\mathbf{y} = (y_1,\ldots,y_m)$.

The range of $\mathbf{u}(t)$, which is independent of time, constitutes the input space of the object. The range of the segmented time function $\mathbf{u} = \mathbf{u}(t_0,t_1)$ is the input segment space. It is important to note that $R[\mathbf{u}]$ depends on the parameters t_0 and t_1 which vary over T. In essence, $\mathcal{Q}$ is associated with a family $R[\mathbf{u}]$ generated by the parameters t_0 and t_1. The same can be said for the range of the output vector $R[\mathbf{y}]$.

In principle the relations between the u_i and y_j can be established by letting **u** vary over its allowable range $R[\mathbf{u}]$ and observing **y**. The input-output relations of (1.3) can be expressed as

$$\begin{aligned} &\mathcal{A}^1(u_1,u_2,\ldots u_k,y_1,y_2,\ldots y_m) = 0 \\ &\ldots\ldots\ldots\ldots\ldots\ldots\ldots\ldots \\ &\mathcal{A}^m(u_1,u_2,\ldots u_k,y_1,y_2,\ldots y_m) = 0 \end{aligned} \tag{1.4}$$

or

$$\begin{aligned} (\mathbf{u},\mathbf{y}) &= 0 \\ \mathbf{y} &= \mathcal{A}(\mathbf{u}) \end{aligned} \tag{1.5}$$

To illustrate equation (1.5) let the physical object $\mathcal{P}$ be an electrical capacitor. A simple experiment is devised in which the voltage v across the capacitor is being measured. Let v be designated as the output and the current i through the capacitor as the input. We have

$$C\frac{dv}{dt} = i$$

where C is a constant. This relation can readily be put in the form of (1.5) as

$$\mathcal{P}(i,v) = 0$$

Note that in the above equations v is not uniquely defined as a function of i. We can determine v to within an added constant (constant of integration). Also, the roles of i and v can be interchanged with no change in formulation or results.

1.4 ORIENTED MODEL

An abstract object $\mathcal{A}$ that admits to (1.5) is an abstract oriented model of $\mathcal{P}$. It can be represented as $\mathcal{A}(\mathcal{P})$. Similarly, $\mathcal{P}$ is a physical realization of $\mathcal{A}$ and can be represented as $\mathcal{P}(\mathcal{A})$.

An abstract oriented model can be formulated to represent every physical object. However, the converse is not always true, i.e., it is possible to generate abstract oriented and nonoriented models for which physical realizations are not possible. (Clearly, an abstract object whose terminal variables, say v_1 and v_2, are represented by a relation of the form $v_1 = jv_2$, where $j = \sqrt{-1}$, cannot be realized.) On the other hand if $\mathcal{A}$ is physically realizable, it can be realized in a variety of physical forms, $\mathcal{P}_1,\mathcal{P}_2 \ldots$. Thus $\mathcal{A}(\mathcal{P}_1),\mathcal{A}(\mathcal{P}_2)$, ... all represent the same abstract object $\mathcal{A}$. To illustrate this point, consider the straight line motion of a mass m accelerating along the OX axis. Let $\mathcal{P}$ represent the physical mass. The variable attributes of $\mathcal{P}$ are chosen as the force F applied along the axis, the position x, velocity $\dot{x}$ and acceleration $\ddot{x}$. Our experiment is to vary F and observe $\ddot{x}$. For the abstract oriented model of $\mathcal{P}$, i.e., $\mathcal{A}(\mathcal{P})$, we designate the input-output as $u(=F)$ and $y(=\ddot{x})$. Thus, the input-output relationship is

$$\mathcal{A}(\mathcal{P}) = \mathcal{A}(F,\ddot{x}) = 0$$

where

$$F = m\ddot{x}$$

The corresponding relation for $\mathcal{P}(\mathcal{A})$ is

$$\mathcal{P}(\mathcal{A}) = \mathcal{P}(u,y) = 0$$

where

$$u = my$$

If we start with $\mathcal{A}$, where $u = my$, then $\mathcal{P}$ is a physical realization of $\mathcal{A}$ with $\ddot{x}$ and F identified as u and y, respectively. Similarly, $\mathcal{A}$ can be realized by a resistor where the current through the resistor is identified with y and the voltage across the resistor is identified with u. Other similar physical examples of $\mathcal{A}$ can easily be identified pointing out the fact that the abstract model $\mathcal{A}$ can be realized in a variety of physical forms $\mathcal{P}$.

To solidify the analytic meaning of an oriented abstract object consider the observation interval $[t_0,t_1]$. Let the vector pair $(\mathbf{u},\mathbf{y})$, where $\mathbf{u} = \mathbf{u}[t_0,t_1]$ and $\mathbf{y} = \mathbf{y}[t_0,t_1]$, be an ordered pair of time functions defined on $[t_0,t_1]$. Let the set $\{(\mathbf{u},\mathbf{y})\}$ be the family of such pairs generated by varying t_0 and t_1 over the time segment T, with $t_1 \geqslant t_0$. Accordingly, an *oriented abstract object* $\mathcal{A}$ is defined[1] as a family of sets of ordered pairs of time functions $\{\{(\mathbf{u},\mathbf{y})\}\}$. The generic pair is $(\mathbf{u},\mathbf{y})$. $\mathbf{u}$ is the input segment and $\mathbf{y}$ is the output segment. $(\mathbf{u},\mathbf{y})$ is the input-output pair belonging to $\mathcal{A}$ if $(\mathbf{u},\mathbf{y})$ is an element of the set $\{(\mathbf{u},\mathbf{y})\}$ for some $[t_0,t_1]$ in T. Thus, an oriented abstract object can be identified with the totality of input-output pairs belonging to $\mathcal{A}$.

To ensure that the family of sets $\{\{(\mathbf{u},\mathbf{y})\}\}$ is defined in a consistent manner, we require that all members of $\{\{(\mathbf{u},\mathbf{y})\}\}$ satisfy the consistency con-

1. *op. cit.*

Note: Notationally { } denotes a set (or family). The double {{ }} denotes a family of sets.

dition. The condition for consistency is that if $(\mathbf{u}[t_0,t_1],\mathbf{y}[t_0,t_1])$ is an input-output pair belonging to $\mathcal{A}$, then any section of the pair within $[t_0,t_1]$ also belongs to $\mathcal{A}$. Specifically, for any pair $(\mathbf{u}[\tau_0,\tau_1],\mathbf{y}[\tau_0,\tau_1])$ where $t_0 \leqslant \tau_0 \leqslant \tau_1$, $\tau_0 \leqslant \tau_1 \leqslant t_1$, we require that $\mathbf{u}[\tau_0,\tau_1] = \mathbf{u}[t_0,t_1]$ and $\mathbf{y}[\tau_0,\tau_1] = \mathbf{y}[t_0,t_1]$ over the interval $[\tau_0,\tau_1]$.

The sets of all segments of $\mathbf{u}$ and $\mathbf{y}$ over (t_0,t_1), such that the pair $(\mathbf{u},\mathbf{y})$ belongs to object $\mathcal{A}$, are referred to as the input segment space $R[\mathbf{u}]$ and output segment space $R[\mathbf{y}]$, respectively. It is implied that the set $\{(\mathbf{u},\mathbf{y})\}$ is a subset of the product space $R[\mathbf{u}] \times R[\mathbf{y}]$. This relationship between $\{(\mathbf{u},\mathbf{y})\}$ and $R[\mathbf{u}] \times R[\mathbf{y}]$ is illustrated in Figure 1.3. Axes aa' and bb' represent $R[\mathbf{u}]$ and $R[\mathbf{y}]$, respectively. The area $ABCD$ represents the product space created by $R[\mathbf{u}] \times R[\mathbf{y}]$. The object area represents $\{(\mathbf{u},\mathbf{y})\}$.

As a simple example of an ordered input-output pair belonging to the abstract object $\mathcal{A}$ consider the mass object of Section 1.4. The dynamical behavior of the mass m under the influence of an external force F is described by the equation

$$F = m\frac{d}{dt}\dot{x}$$

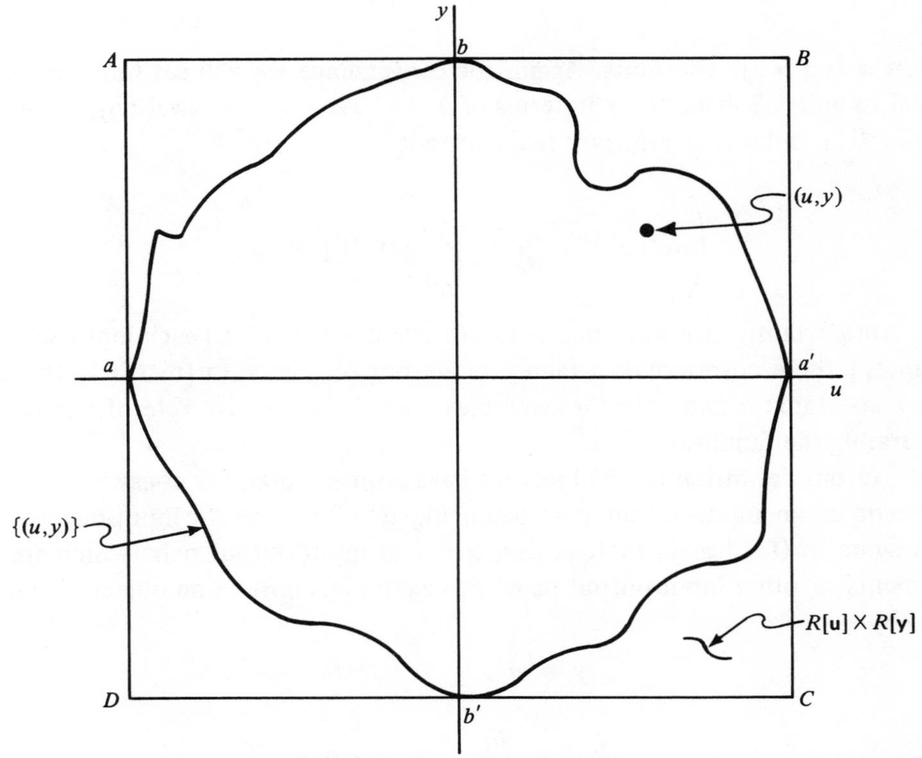

Figure 1.3 Representation of an abstract object.

Designating the input and output, respectively, as $u = F$ and $y = \dot{x}$ we have from equation (1.5)

$$\mathfrak{A}(u,y) = 0$$

$$u = m\frac{dy}{dt}$$

The generic input-output pair (u,y) over the finite time interval (t_0,t_1) is the pair of time functions

$$\left(u(t), a + \frac{1}{m}\int_{t_0}^{t} u(\xi)d\xi\right)$$

where $t_0 \leqslant t \leqslant t_1$ and a is a real arbitrary constant (of integration). Such pairs as $(1,t), (1,1+t), (1,2+t), \ldots, (t, 1+t^2/2), (t, 2+t^2/2), \ldots,$ represent a family of sets of input-output pairs for the mass object. (For convenience the mass m was normalized to unity.)

As another example of an oriented abstract object consider a simple feedback control system characterized by the dynamic relation

$$\frac{dy}{dt} = ay + bu$$

where a and b are constants. Again, for convenience we will set both constants equal to unity. Solving for y in terms of u, we have for (u,y) over (t_0,t_1), where $t_0 \leqslant t \leqslant t_1$ and α is an arbitrary real constant,

$$\left(u(t), e^{-(t-t_0)}\alpha + \int_{t_0}^{t_1} e^{-(t-\xi)}u(\xi)d\xi\right)$$

Importantly, we note that, in both examples cited, to each input segment $u[t_0,t_1]$ there corresponds a family of output segments $\{y[t_0,t_1]\}$. The arbitrary constants a and α in the examples play the respective role of parameters generating the families.

In our definition of an object we have required that, for consistency, every *segment* of an input-output pair belonging to $\mathfrak{A}$ also be an input-output pair belonging to $\mathfrak{A}$. Clearly, there are segments of input-output pairs which are not segments of other input-output pairs. For example, consider an object characterized by

$$y = u \qquad t \geqslant t_0,\ t_0 < 0$$

$$\frac{dy}{dt} = \frac{du}{dt} \qquad t \geqslant 0$$

For $t \geqslant 0$ input-output pairs such as (1,3), (4,5), (6,1), ..., where u and y are constants, all satisfy the characterization relationship $\dot{y} = \dot{u}$. However, they are not pairs defined on the interval $t \geqslant t_0$, $t_0 < 0$, where $(u(t), u(t))$ is the generic form. It is for this reason we define $\mathcal{A}$ as a family of sets of input-output pairs.

In the applications to follow it will be sufficient to restrict our attention to objects characterized by a single set of input-output pairs defined over the interval T. Such objects are said to be *uniform*. Specifically, an object $\mathcal{A}$ is said to be a uniform oriented object if every input-output pair $(\mathbf{u}[t_0,t_1], \mathbf{y}[t_0,t_1])$ belonging to $\mathcal{A}$ is a segment of an input-output pair $(\mathbf{u}_T, \mathbf{y}_T)$ defined over T. Thus a uniform oriented object can be characterized by a single set of pairs $\{(\mathbf{u}_T, \mathbf{y}_T)\}$. If, for example, $T = [0,\infty)$ then $\mathcal{A}$ is characterized by the single set of ordered pairs $\{(\mathbf{u}[0,\infty), \mathbf{y}[0,\infty))\}$ defined over T. For convenience, and unless otherwise stated in the sequel, we will assume that $\mathcal{A}$ is a uniform oriented object, and that $T = (-\infty,\infty)$.

1.5 NOTION OF STATE

The abstract object has been defined as a *relation*, i.e., a set of ordered pairs $(\mathbf{u},\mathbf{y})$ rather than a function where for each $\mathbf{u}$ there corresponds a unique $\mathbf{y}$. To a given $\mathbf{u}$ there is associated a *set* of output $\mathbf{y}$'s. $\mathbf{u}$ and each $\mathbf{y}$ in the set comprise an input-output pair belonging to $\mathcal{A}$. Thus, we have departed from the conventional definitions which identify $\mathcal{A}$ within a function (or an operator) and associate a unique output for each input.

The set of distinct $\mathbf{y}$'s associated with a given input $\mathbf{u}$ is generated by the different initial conditions under which $\mathbf{u}$ influences $\mathcal{A}$. (These different initial conditions are equivalent to "initial states.") One way of associating a *unique* $\mathbf{y}$ with each $\mathbf{u}$ is to attach a distinguishing lable to each pair $(\mathbf{u},\mathbf{y})$. In the discussion to follow this label, which we denote as $\mathbf{x}(t_0)$, will be called the *state* of $\mathcal{A}$ at time t_0. The state will span a (state) space, which we will denote as Ψ, in such a way that $\mathbf{y}$ will be uniquely determined by $\mathbf{u}$ and $\mathbf{x}(t_0)$. The process of attaching the state label is called *parametrization* of the space of input-output pairs. Essentially, by establishing the state of $\mathcal{A}$ at time t_0 we separate the past from the future. We provide that information about the past that is relevant to determining the response of $\mathcal{A}$ beginning at time t_0.

1.6 STATE-INPUT-OUTPUT RELATIONS

There are many ways in which a state vector $\mathbf{x}(t_0)$ can be associated with $\mathcal{A}$. Therefore the definitions given below will be a set of qualifying conditions which, if met, allow one to label the state of $\mathcal{A}$ at time t_0 as $\mathbf{x}(t_0)$.

For discussion purposes consider the semi-closed interval $(t_0,t]$ and the ordered pair $(\mathbf{u}(t_0,t], \mathbf{y}(t_0,t])$, which is a segment of $(\mathbf{u}_T, \mathbf{y}_T)$ defined over T. Let $\mathcal{A}$ be characterized over $(t_0,t]$ by input-output relation (1.5), i.e., by

$$\mathcal{A}(\mathbf{u},\mathbf{y}) = 0$$
$$\mathbf{y} = \mathcal{A}(\mathbf{u})$$

(This is equivalent to saying $\mathcal{A}$ is characterized by the *family* of *sets* of ordered pairs $\{\{(\mathbf{u},\mathbf{y})\}\}$. Let $\mathbf{x}$ be a (vector) variable ranging over space Ψ. It is claimed[1] that if the spaces of input-output pairs admit to a relation of the form

$$\mathbf{y}(t_0,t] = \mathbf{F}(\mathbf{x};\mathbf{u}(t_0,t]) \tag{1.6}$$

which satisfies the four mutual- and self-consistency conditions set forth below, then (a) equation (1.6) qualifies as an *input-output-state* relation for $\mathcal{A}$, (b) space Ψ qualifies as the state-space for $\mathcal{A}$, where elements of Ψ are the states, and (c) the variable $\mathbf{x}_0 = \mathbf{x}(t_0)$ qualifies as the state of $\mathcal{A}$ at time t_0. Thus, if the consistency conditions as defined below are satisfied it can be said that $\mathcal{A}$ is completely characterized by (1.6), where $\mathbf{y}(t_0,t]$ is the response segment of $\mathcal{A}$ to input segment $\mathbf{u}(t_0,t]$ starting in state $\mathbf{x}_0$ and $(\mathbf{u}(t_0,t], \mathbf{y}(t_0,t])$ is the input-output pair associated with $\mathbf{x}_0$.

The conditions of mutual- and self-consistency which qualify (1.6) as the input-output-state relation for $\mathcal{A}$ are defined[2] below.

1.7 CONDITION I, MUTUAL-CONSISTENCY

Every input-output pair for $\mathcal{A}$ satisfies relation (1.6), and conversely. To elaborate, if $(\mathbf{u}(t_0,t], \mathbf{y}(t_0,t])$, or simply $(\mathbf{u},\mathbf{y})$, satisfies (1.5) then the pair also satisfies (1.6) since there exists an $\mathbf{x}$ (say $\mathbf{x}_0$) in Ψ such that

$$\mathbf{y}(t_0,t] = \mathbf{F}(\mathbf{x}_0;\mathbf{u}(t_0,t]) \tag{1.7}$$

Conversely, any pair $(\mathbf{u},\mathbf{y})$ satisfying (1.6) for some $\mathbf{x}$ in Ψ over $(t_0,t]$ is also an input-output pair for $\mathcal{A}$. This condition must hold for all $(t_0,t]$ in T and for all $\mathbf{u}$ in $R[\mathbf{u}]$. The purpose of the mutual-consistency condition is to ensure that equations (1.5) and (1.6) both represent the same object.

We demonstrate the mutual-consistency condition by again considering the mass object of Section 1.4. $\mathcal{A}$ is characterized by

1. *op. cit.*
2. *op. cit.*

$$\frac{dy}{dt} = u \qquad (m = 1) \tag{1.8}$$

The first part of the mutual-consistency condition is fulfilled in the sense that (1.8) can be written as

$$\mathfrak{A}(\mathbf{u},\mathbf{y}) = 0$$

where

$$(u,y) = \left(u(t), a + \int_{t_0}^{t} u(\xi)\,d\xi\right)$$

thereby satisfying (1.5). Further, the output y satisfies (1.6) since we can write, with proper choice of x,

$$y(t) = x_0 + \int_{t_0}^{t} u(\xi)\,d\xi$$

$$= F(x_0;u(t))$$

Conversely, consider the input-output state relation of the form

$$y(t) = \alpha + \int_{t_0}^{t} u(\xi)\,d\xi$$

$$= F(\alpha;u(t)) \tag{1.9}$$

With proper choice of x $(=\alpha)$ the pair $(u(t),y(t))$ characterizing $\mathfrak{A}$ satisfies (1.6). By direct substitution of (1.9) into (1.8) it follows that every pair (u,y) satisfying (1.9) also satisfies (1.8). Thus, every input-output pair for $\mathfrak{A}$ characterized by (1.8) satisfies (1.6), and vice versa.

1.8 CONDITION II, FIRST SELF-CONSISTENCY

The response $\mathbf{y}(t)$ at any time $t > t_0$ is uniquely determined by $\mathbf{x}$ and $\mathbf{u}(t_0,t]$. This must be true for all t_0. To qualify as a state space of $\mathfrak{A}$, the space Ψ must have the property that, given any point $\mathbf{x}$ in Ψ and any input $\mathbf{u}(t_0,t]$ (defined over the input segment space), the output at time t is uniquely determined by $\mathbf{x}$

and **u**. The point **x** at time t_0 will be called the state of $\mathcal{A}$ at t_0. The output is independent of **u** or **y** prior to time t_0. This is a key property of the state space concept.

An input-output-state relation of form (1.9) clearly satisfies the first self-consistency condition—regardless of the range of α. (For all t_0 the output y at time t is uniquely determined by α and u. y is independent of u or y prior to t_0.) However, by changing the upper limit of the integral (1.9) from t to $t+1$, i.e., if we have

$$y(t) = \alpha + \int_{t_0}^{t+1} u(\xi)d\xi \qquad t > t_0$$

we no longer satisfy the first self-consistency condition. The output at time t cannot be determined without knowledge of $u(\xi)$ between t and $t+1$.

1.9 CONDITION III, SECOND SELF-CONSISTENCY

If the input-output pair $(\mathbf{u}[t_0,t_1], \mathbf{y}[t_0,t_1])$ satisfies (1.6), then $(\mathbf{u}[t,t_1], \mathbf{y}[t,t_1])$ also satisfies (1.7). $\mathbf{u}[t,t_1]$ and $\mathbf{y}[t,t_1]$ are sections of $\mathbf{u}[t_0,t_1]$ and $\mathbf{y}[t_0,t_1]$, respectively, and $t_0 \leqslant t \leqslant t_1$. This must hold for all **x** in Ψ, all $\mathbf{u}[t_0,t_1]$ in the input segment space, and all t_0,t,t_1. The purpose of this condition is to ensure that the state space Ψ can include all possible initial conditions for $\mathcal{A}$.

To help clarify the meaning of the second self-consistency condition consider the input to $\mathcal{A}$ over $(t_0,t_1]$ as consisting of two contiguous segments, $\mathbf{u}^0(t_0\, t]$ *followed* by $\mathbf{u}^1(t,t_1]$, i.e.,

$$\mathbf{u} = \mathbf{u}^0\mathbf{u}^1$$

(Note: $\mathbf{u}^0\mathbf{u}^1$ is *not* to be interpreted as the product of $\mathbf{u}^0$ and $\mathbf{u}^1$.) If the input-output pair $(\mathbf{u}^0\mathbf{u}^1,\mathbf{y}^0\mathbf{y}^1)$ satisfies (1.6), then the segment response to $\mathbf{u}^0\mathbf{u}^1$ beginning in state $\mathbf{x}(t_0) = \mathbf{x}_0$ is

$$\mathbf{y}^0\mathbf{y}^1 = \mathbf{F}(\mathbf{x}_0;\mathbf{u}^0\mathbf{u}^1) \tag{1.10}$$

To say that $(\mathbf{u}^1,\mathbf{y}^1)$ also satisfies (1.6), as the second self-consistency condition does, means that there must be some values of **x** in Ψ such that

$$\mathbf{y}^1 = \mathbf{F}(\mathbf{x};\mathbf{u}^1) \tag{1.11}$$

for all arbitrary times t, where $t_0 < t \leqslant t_1$. Denoting these values of **x** in Ψ as the set A the second self-consistency condition requires that this set be non-empty. (Figure 1.4 graphically illustrates the idea.)

To further illustrate the second self-consistency condition let $\mathcal{A}$ be characterized by the relation

$$y(\xi) = \alpha^2 + \int_{t_0}^{t_1} u(\xi)\,d\xi \tag{1.12}$$

u and y are real-valued time functions and Ψ includes the set of all real numbers, i.e., $\alpha = x$ is any real number. Let's consider the value of y at the particular time $\xi = t$. For $\xi = t$ equation (1.12) becomes

$$y(t) = \alpha^2 + \int_{t_0}^{t_1} u(t)\,dt \tag{1.13}$$

Let the input $u(t)$ over the observation interval $[0,5]$ be $u(t) = -3t^2$ and let the initial conditions at time $t_0 = 0$ be $x_0 = \alpha_0 = 1$. From (1.13) the response over $[0,5]$ is

$$y(t) = 1 - t^3 \Big|_0^5$$

Thus, the input-output pair associated with $x_0 = 1$ for the interval $[0,5]$ is $(-3t^2, 1 - t^3)$.

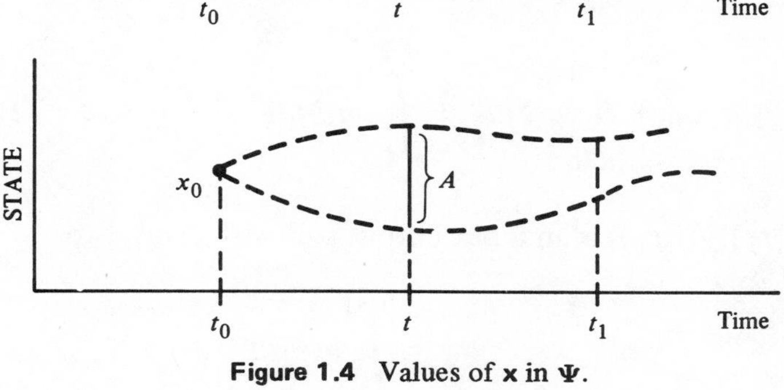

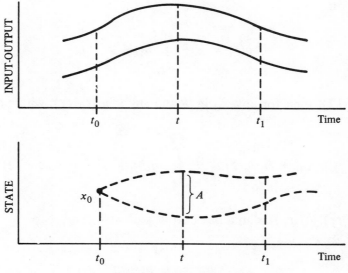

Figure 1.4 Values of $\mathbf{x}$ in Ψ.

Now consider the intermediate time $t = 2$. The second self-consistency condition requires that if the input-output pair $(u(0,5], y(0,5])$, which is $(-3t^2, 1 - t^3)$, satisfies (1.6), then the same pair over the time interval [2,5] also must satisfy (1.6). Since $\mathcal{Q}$ is characterized by (1.12), we have for [2,5],

$$y(t) = \alpha_1^2 + \int_2^5 -3t^2 dt \tag{1.14}$$

Also, from (1.12) we can write for the pair $(-3t^2, 1 - t^3)$

$$y(t) = 1 + \int_0^2 -3t^2 dt + \int_2^5 -3t^2 dt \tag{1.15}$$

The only way the pair $(-3t^2, 1 - t^3)$ can satisfy (1.14) is if

$$\alpha^2 = 1 - \int_0^2 3t^2 dt$$

$$= -7$$

Clearly, $\alpha^2 = -7$ has no solution in Ψ, where Ψ consists of real numbers. Therefore A is empty and the second self-consistency condition is not satisfied by relation (1.12).

On the other hand, if $\mathcal{Q}$ were characterized by an input-output-state relation of the form

$$y(t) = \alpha + \int_{t_0}^{t} u(\xi) d\xi \tag{1.16}$$

instead of (1.12), then for any $\mathbf{x}_0$ in Ψ and any ξ in $[t_0, t_1]$, where $t_0 < \xi \leqslant t_1$, we can write

$$y(\xi) = \alpha_0 + \int_{t_0}^{t} u(\xi) d\xi + \int_{t}^{t_1} u(\xi) d\xi \qquad t_0 < t < t_1 \tag{1.17}$$

The pair $(u(t,t_1), y(t,t_1))$ is an input-output pair with respect to

$$x_1 = \alpha_0 + \int_{t_0}^{t} u(\xi) d\xi \tag{1.18}$$

It is seen from (1.18) that for any real-valued time function u in the input space the state x_1 assumes a real value in Ψ, given that α_0 is real. Thus an input-output-state relation of form (1.16) satisfies the second self-consistency condition.

1.10 CONDITION IV, THIRD SELF-CONSISTENCY

We defined an object $\mathcal{A}$ as a family of sets of input-output pairs. To complete the parametrization of the input-output pairs it remains to examine the meaning of the term "the state of $\mathcal{A}$ at time t," where t is arbitrary. This is equivalent to asking what happens to the set A of values of $\mathbf{x}$ in Ψ as the input segment $\mathbf{u}^1(t,t_1]$ varies over the input segment space of $\mathcal{A}$. We want to portray the situation as one where $\mathbf{x}_0$ and $\mathbf{u}^0$ are held fixed and $\mathbf{u}^1$ is varied over the entire input segment space.

From the first and second self-consistency conditions we readily deduce that the set A depends on $\mathbf{x}_0$, $\mathbf{u}^0$ and $\mathbf{u}^1$ (where the variable t is arbitrarily fixed). We can write for A

$$A = A(\mathbf{x}_0;\mathbf{u}^0\mathbf{u}^1) \tag{1.19}$$

Now consider the situation where $\mathbf{u}^1$ varies over the input segment space of $\mathcal{A}$, i.e., $\mathbf{u}^1$ assumes the variations $\mathbf{u}_1^1,\mathbf{u}_2^1,\ldots$. Accordingly, the corresponding A sets become $A(\mathbf{x}_0;\mathbf{u}^0\mathbf{u}_1^1), A(\mathbf{x}_0;\mathbf{u}^0\mathbf{u}_2^1),\ldots$ If we form the *intersection* of all the A sets (denoted by the A' subset) this intersection (if not empty) is the set of all points $\mathbf{x}$ in Ψ to which every pair $(\mathbf{u}^1,\mathbf{y}^1)$ satisfying

$$\mathbf{y}^0\mathbf{y}^1 = \mathbf{F}(\mathbf{x}_0;\mathbf{u}^0\mathbf{u}^1) \tag{1.20}$$

also satisfies

$$\mathbf{y}^1 = \mathbf{F}(\mathbf{x};\mathbf{u}^1) \tag{1.21}$$

for all $\mathbf{x}$ in the intersection.

Based on the above discussion we state the third self-consistency condition as follows: Let $(\mathbf{u}^0\mathbf{u}^1,\mathbf{y}^0\mathbf{y}^1)$ be an input-output pair satisfying (1.6) with respect to some $\mathbf{x}_0$, i.e.,

$$\mathbf{y}^0\mathbf{y}^1 = \mathbf{F}(\mathbf{x}_0;\mathbf{u}^0\mathbf{u}^1)$$

Also, let $A(\mathbf{x}_0;\mathbf{u}^0\mathbf{u}^1)$ be the set of all $\mathbf{x}$ where $(\mathbf{u}^1,\mathbf{y}^1)$ is an input-output pair satisfying (1.7), i.e.,

$$\mathbf{y}^1 = \mathbf{F}(\mathbf{x};\mathbf{u}^1)$$

The third self-consistency condition requires that the *intersection* of all $A(\mathbf{x}_0;\mathbf{u}^0\mathbf{u}^1)$ taken over all $\mathbf{u}^1(t,t_1]$ in the input segment space be nonempty for all t_0, all $\mathbf{x}_0$ in Ψ, and all segments $\mathbf{u}(t_0,t_1]$ in the input segment space of $\mathcal{A}$.

From the way in which the third self-consistency condition is formulated it cannot be satisfied unless *all* of the $A(\mathbf{x}_0;\mathbf{u}^0\mathbf{u}^1)$ sets are nonempty. This implies that the second self-consistency condition is also satisfied. However, the converse is not true.

The meaning of the third self-consistency condition is readily demonstrated by example (1.16). Let $\mathcal{A}$ be characterized by an input-output relation of the form

$$y(\xi) = x_0 + \int_{t_0}^{t_1} u(\xi)d\xi \tag{1.22}$$

We establish the following conditions under which the third self-consistency condition will be evaluated: $t_0 = 0$, $x_0 = 4$, intermediate time $t = 2$, $t_1 > 2$, $u(\xi) = 2\xi$ for $0 < \xi \leqslant t$, and $u(\xi) = $ variable for $2 < \xi \leqslant t_1$. Dividing the integration range of (1.22) into [0,2] and $[2,a]$, where $2 < a \leqslant t$, we have

$$y(a) = 4 + \int_0^2 2\xi d\xi + \int_2^a u(\xi)d\xi \qquad 2 < a \leqslant t$$

$$= 8 + \int_2^a u(\xi)d\xi \tag{1.23}$$

which is of the same form as (1.22). This implies that $(u(2,t_1],y(2,t_1])$, with $y(2,t_1]$ defined by (1.23), satisfies (1.22) with $x = 8$. Therefore, the A sets as defined by (1.19), (1.20) and (1.21) are nonempty. These sets all contain the point $x = 8$, and hence, so does their intersection. We conclude A is nonempty when $t = 2$, $u(\xi) = 2\xi$ for $0 < \xi \leqslant t$, and that relation (1.22) satisfies the third self-consistency condition.

1.11 STATE AT TIME t

For the discussion to follow we denote the intersection of all the A sets (formed by varying $\mathbf{u}^1$ over the entire input segment space) as the A' subset. (It has been said that A' is the set of all points $\mathbf{x}$ in Ψ to which every pair $(\mathbf{u}^1,\mathbf{y}^1)$ satisfying (1.20) also satisfies (1.21).) Clearly, A' contains values of $\mathbf{x}$ where the response to $\mathbf{u}^1$ is the same for all starting states in A'. This is true for all $\mathbf{u}^1$ in the input segment space. Therefore, the third self-consistency condition provides for the following definition[1] of the *state of $\mathcal{A}$ at time t*: Given that $\mathcal{A}$ is characterized by (1.6) and the response of $\mathcal{A}$ to input segment $\mathbf{u}^0\mathbf{u}^1$ starting in state $\mathbf{x}_0$ is described by (1.20) and (1.21), then the state of $\mathcal{A}$ at time t is any state $\mathbf{x}$ in the A' subset.

Symbolically, the state of $\mathcal{A}$ at time t will be denoted as $\mathbf{x}(t)$, bearing in mind that this is merely a label to identify an element of $A'(\mathbf{x}_0;\mathbf{u}^0\mathbf{u}^1)$ with the state at time t. Similarly, $\mathbf{x}(t_0)$ will notationally identify with the initial state of $\mathcal{A}$ at time t_0. Accordingly, the response at time t can be written, in accordance with (1.6), as

$$\mathbf{y}(t) = \mathbf{y}(\mathbf{x}(t_0);\mathbf{u}(t_0,t]) \tag{1.24}$$

We note from (1.24) that, for each fixed t, $\mathbf{x}(t_0)$ and $\mathbf{x}(t)$ range over the entire state space Ψ, i.e., $R[\mathbf{x}(t)] = \Psi$. The components of $\mathbf{x}(t)$ will therefore be designated as the state variables or *elements of the state vector* $\mathbf{x}(t)$.

On considering an input $\mathbf{u}(t_0,t_1]$ applied to $\mathcal{A}$ while $\mathcal{A}$ is in state $\mathbf{x}(t_0)$, it follows that $\mathbf{x}(t_1)$ is the terminal state. If, however, we have $\mathbf{u}(t_0,t_1]$ followed by $\mathbf{u}(t_1,t_2],\mathbf{u}(t_2,t_3],\ldots,$ then each of the respective terminal states $\mathbf{x}(t_1),\mathbf{x}(t_2),\mathbf{x}(t_3),\ldots,$ plays a dual role; each represents both the terminal state relative to input segment t and the input state relative to input segment $t+1$.

For objects characterized by input-output relations of the form (1.5), i.e., $\mathcal{A}(\mathbf{u},\mathbf{y}) = 0$, the standard practice is to associate $\mathbf{x}(t)$ with such objects by defining $\mathbf{x}(t)$ in terms of $\mathbf{u}(t_0,t]$ and $\mathbf{y}(t_0,t]$, and to verify that the four consistency conditions are satisfied. In cases where (1.5) is a differential equation then $\mathbf{x}(t)$ can be defined in terms of $\mathbf{u}(t)$ and $\mathbf{y}(t)$ and a finite number of their time derivatives. Other expressions for $\mathbf{x}(t)$ in terms of $\mathbf{u}(t_0,t]$ and $\mathbf{y}(t_0,t]$ can readily be obtained by changing the coordinate system of the state space. Importantly, there are many ways to associate a state vector with an object.

$\mathbf{x}(t)$ has been defined to be an element of the set $A'(\mathbf{x}(t_0);\mathbf{u}^0\mathbf{u}^1)$. Since A' depends only on $\mathbf{x}(t_0)$ and $\mathbf{u}(t_0,t]$, then so must $\mathbf{x}(t)$. Thus, we can write for the *state* equation

$$\mathbf{x}(t) = \mathbf{x}(\mathbf{x}(t_0);\mathbf{u}(t_0,t]) \tag{1.25}$$

Under proper regularity assumptions for $\mathbf{x}$ and $\mathbf{u}$ equations (1.24) and (1.25) will tend in the limit to differential equations of the form

$$\dot{\mathbf{x}}(t) = \mathbf{x}(\mathbf{x}(t),\mathbf{u}(t),\ldots,\mathbf{u}^{(k)}(t),t) \tag{1.26}$$

$$\mathbf{y}(t) = \mathbf{y}(\mathbf{y}(t),\mathbf{u}(t),\ldots,\mathbf{u}^{(k)}(t),t) \tag{1.27}$$

1. *op. cit.*

where $\mathbf{x}$ and $\mathbf{y}$ are point functions and $\mathbf{u}^{(k)}(t)$ is the kth derivative of $\mathbf{u}(t)$. Equations (1.26) and (1.27) particularly apply to differential objects and are referred to as the *state equations* in *differential form*. Differential objects and their respective state equations are of considerable interest in system theory and will be the focus of discussion in later chapters. In particular, linear differential systems and their state equations assume the simple form of

$$\dot{\mathbf{x}}(t) = \mathbf{A}(t)\mathbf{x}(t) + \mathbf{B}(t)\mathbf{u}(t) \tag{1.28}$$

$$\mathbf{y}(t) = \mathbf{C}(t)\mathbf{x}(t) + \mathbf{D}(t)\mathbf{u}(t) + \mathbf{D}_1(t)\mathbf{u}^{(1)}(t) + \ldots + \mathbf{D}_k(t)\mathbf{u}^{(k)}(t) \tag{1.29}$$

where the coefficients are (time dependent) matrices.

1.12 MULTIPLE OBJECTS

Thus far we have considered only the single object $\mathcal{A}$, its terminal variables and the relations between them. Our principal interest, however, is a collection of objects that interact with one another. We will now focus on the analysis of multiple objects.

For the most part we shall use the conventional rectangular block to graphically describe an abstract object $\mathcal{A}$ (see Figure (1.5)). Here the leads represent the terminal variables, which can denote either scalar or vector variables. The important thing is that each variable is treated as an entity when each object $\mathcal{A}$ is interconnected with other objects.

From equation (1.25), where

$$\mathbf{y}(t) = \mathbf{y}(\mathbf{x};\mathbf{u})$$

it is seen that each component of $\mathbf{y}(t)$ is a function of $\mathbf{x}$ and $\mathbf{u}$. Therefore, we can express each component of $\mathbf{y}$ as

$$y_i(t) = y_i(\mathbf{x};\mathbf{u}) \tag{1.30}$$

This implies that an object, as shown in Figure 1.5, having ℓ inputs and m outputs, can be represented by m objects $\mathcal{A}_1, \mathcal{A}_2, \mathcal{A}_3, \ldots, \mathcal{A}_m$ each having ℓ

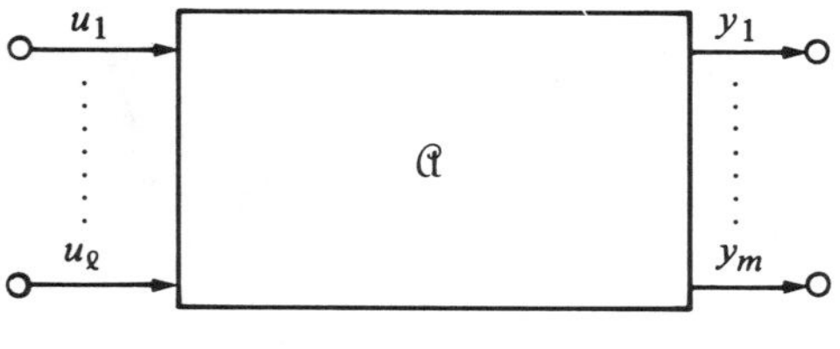

Figure 1.5 Graphical representation of object.

inputs and one output (see Figure 1.6). Thus, if the objects under consideration are oriented there is no loss in generality in assuming that the composite "object" is made up of oriented objects where each has one output variable.

1.13 INTERCONNECTED OBJECTS

Essentially, a "system" is a collection of interacting objects. The interactions between objects $\mathcal{A}_1, \mathcal{A}_2, \ldots, \mathcal{A}_N$, represent constraints on the terminal variables. An example of such a constraint is the fact that the ith terminal variable of object $\mathcal{A}_j$ is equal to the kth terminal variable of object $\mathcal{A}_\ell$, for all t in T. The set of objects $\mathcal{A}_1, \mathcal{A}_2, \ldots, \mathcal{A}_N$, constrained in this manner is said to be an interconnection of $\mathcal{A}_1, \mathcal{A}_2, \ldots, \mathcal{A}_N$. Such an interconnection can be represented graphically as shown in Figure 1.7, where the composite object is designated as $\mathcal{A}$. Specifically, the set of objects $\{\mathcal{A}_i\}$ $i = 1,2,3,\ldots,N$, is said to be an interconnection of the $\mathcal{A}_i$ if every object in the set shares at least one terminal variable with one or more of the other objects of the set. If some, or possibly none, of the $\mathcal{A}_i$ share terminal variables the collection is partially interconnected.

There are several simple types of interconnections encountered frequently in systems analysis; they are the tandem combination and the parallel combination. In the sequel we will briefly highlight both.

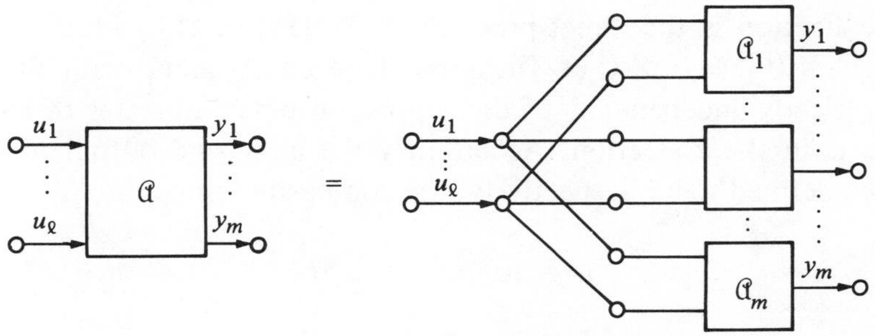

Figure 1.6 Equivalent representation.

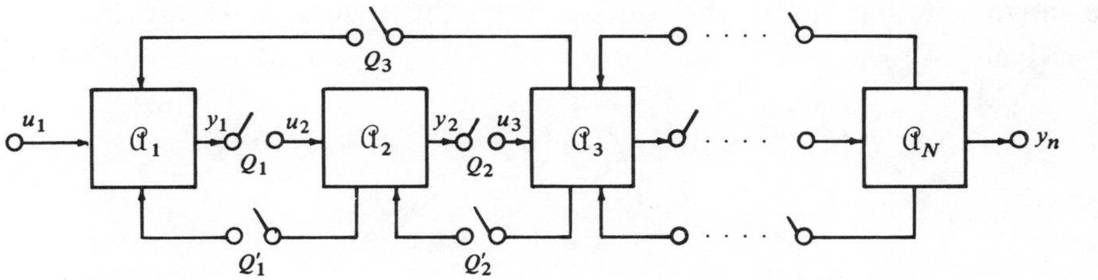

Figure 1.7 Interconnected objects.

1.14 TANDEN COMBINATIONS—INITIALLY FREE

For an analysis on multiple objects to have any concreteness it must be able to answer for the composite object the same questions that are asked of the single object: What is the input-output-state relationship of the composite interconnection? What is the state-space? What is the output function space? At this point it must be noted that answers to these questions, particularly for large complex systems, have not been readily established. In fact some of the concepts suggested below are tentative. The ideas presented below, although tentative and sketchy, are representative of some progress in this area.

In order to establish a basis for answering the above questions we introduce the idea of the *initially free interconnection*, which will be symbolically denoted as $\mathcal{Q}(t_0)$. The connections between the $\mathcal{Q}_i$, as shown in Figure 1.7, remain open until $t = t_0$, at which time the Q switches are closed. We regard $\mathcal{Q}$ as a limiting form (as $t_0 \to -\infty$) of an initially free interconnection $\mathcal{Q}(t_0)$. It is therefore possible to relate the properties of $\mathcal{Q}(t)$ to those of $\mathcal{Q}(t_0)$, thereby giving a characterization to the state space of $\mathcal{Q}$.

Consider a collection of oriented objects $\mathcal{Q}_1, \mathcal{Q}_2, \ldots \mathcal{Q}_N$, (as depicted in Figure 1.7) where each is characterized by a relation of the form

$$\mathbf{y}^i = \mathbf{y}(\mathbf{x}^i;\mathbf{u}^i) \tag{1.31}$$

If there is no interaction between the various $\mathcal{Q}_i$, i.e., all the switches are open, then the collection is the direct product of $\mathcal{Q}_1, \mathcal{Q}_2, \ldots, \mathcal{Q}_N$, which will be denoted as $\mathcal{Q}_1 \times \mathcal{Q}_2 \times \ldots \times \mathcal{Q}_N$. The input to each $\mathcal{Q}_i$ along with the state of each $\mathcal{Q}_i$ is clearly independent of the inputs, outputs and states of any of the other objects in the collection. Accordingly the input and output of the composite $\mathcal{Q}$ are defined to be, respectively, the composite vectors.

$$\mathbf{u} = (\mathbf{u}^1,\mathbf{u}^2,\ldots,\mathbf{u}^N) \tag{1.32}$$

and

$$\mathbf{y} = (\mathbf{y}^1,\mathbf{y}^2,\ldots,\mathbf{y}^N) \tag{1.33}$$

The corresponding input and output segment spaces of $\mathcal{Q}$ are the products, respectively,

$$R[\mathbf{u}] = R[\mathbf{u}^1] \times R[\mathbf{u}^2] \times \ldots \times R[\mathbf{u}^N] \tag{1.34}$$

and

$$R[\mathbf{y}] = R[\mathbf{y}^1] \times R[\mathbf{y}^2] \times \ldots \times R[\mathbf{y}^N] \tag{1.35}$$

Similarly, the composite vector

$$\mathbf{x}(t) = (\mathbf{x}^1(t),\mathbf{x}^2(t),\ldots,\mathbf{x}^N(t)) \tag{1.36}$$

is the state of $\mathcal{A}$ at time t, where the state space of $\mathcal{A}$ is the product space

$$\Psi = \Psi^1 \times \Psi^2 \times \ldots \times \Psi^N \tag{1.37}$$

Now let two of the $\mathcal{A}_i$, say $\mathcal{A}_1$ and $\mathcal{A}_2$, be connected in tandem, with $\mathcal{A}_1$ preceeding $\mathcal{A}_2$, as shown in Figure 1.8. In a tandem combination the input $\mathbf{u}^2$ to $\mathcal{A}_2$ is "constrained" to be equal to the output $\mathbf{y}^1$ of $\mathcal{A}_1$. It is necessary now to distinguish between two situations: (1) $\mathbf{u}^2 = \mathbf{y}^1$ for $t > t_0$, where t_0 is the instant $\mathbf{u}^1$ is applied to $\mathcal{A}_1$, and (2) $\mathbf{u}^2 = \mathbf{y}^1$ for all t. Case (1) is the *initially free* tandem combination, which will be considered subsequently. Case (2) is the *constrained* tandem combination and will be considered in the next section. In (1) $\mathcal{A}_1$ and $\mathcal{A}_2$ act like the direct product for $t < t_0$. The switch Q signifies that $\mathbf{u}^2$ is constrained to be equal to $\mathbf{y}^1$ from time t_0 on, *but not before*.

Proceeding with case (1), where Q is closed at time t_0, we assume that $\mathcal{A}_1$ and $\mathcal{A}_2$ are characterized by relations of form (1.31), with the output segment space of $\mathcal{A}_1$ contained in the input segment space of $\mathcal{A}_2$. Let the states of $\mathcal{A}_1$ and $\mathcal{A}_2$ at time t be (as deduced from (1.31))

$$\mathbf{x}^i(t) = \mathbf{x}(\mathbf{x}_0^i;\mathbf{u}^i) \tag{1.38}$$

where $\mathbf{x}_0^i = \mathbf{x}^i(t_0)$ and $\mathbf{u}^i = \mathbf{u}^i(t_0,t]$. The initial states $\mathbf{x}_0^1$ and $\mathbf{x}_0^2$ can be chosen arbitrarily in Ψ^1 and Ψ^2. Therefore given $\mathbf{x}_0^1$, $\mathbf{x}_0^2$ and $\mathbf{u}^1(t_0,t]$ we can find, from (1.31) and (1.38), the output $\mathbf{y}^2(t_0,t]$, and the states $\mathbf{x}^1(t)$ and $\mathbf{x}^2(t)$ for any $t > t_0$. Thus, the *free product* of the tandem combination $\mathcal{A}_1$ and $\mathcal{A}_2$ is an object $\mathcal{A}(t_0)$ characterized by the input-output-state relation

$$\begin{aligned} \mathbf{y}^2 &= \mathbf{y}[\mathbf{x}_0^2;\mathbf{u}^2] \\ &= \mathbf{y}[\mathbf{x}_0^2;\mathbf{y}(\mathbf{x}_0^1;\mathbf{u}^1)] \end{aligned} \tag{1.39}$$

where the composite state vector

$$\mathbf{x}_0 = (\mathbf{x}_0^1,\mathbf{x}_0^2) \tag{1.40}$$

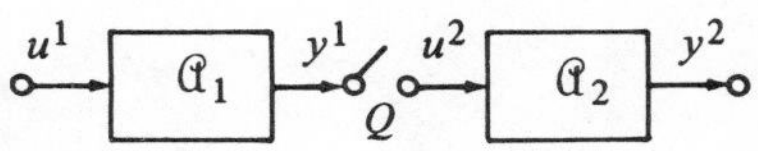

Figure 1.8 Tandem combination of $\mathcal{A}_1$ and $\mathcal{A}_2$.

plays the role of the initial state of $\mathcal{A}(t_0)$. The state of $\mathcal{A}(t_0)$ at time t is

$$\mathbf{x}(t) = (\mathbf{x}^1(t), \mathbf{x}^2(t)) \tag{1.41}$$

where $\mathbf{x}^1(t)$ and $\mathbf{x}^2(t)$ are expressed by (1.38) as

$$\mathbf{x}^1(t) = \mathbf{x}(\mathbf{x}_0^1; \mathbf{u}^1) \tag{1.42}$$

$$\mathbf{x}^2(t) = \mathbf{x}(\mathbf{x}_0^2; \mathbf{x}(\mathbf{x}_0^1; \mathbf{u}^1)) \tag{1.43}$$

The corresponding output segment space for $\mathcal{A}(t_0)$ is the set of output functions $\mathbf{y}^2$, i.e.,

$$R[\mathbf{y}^2] = \{\mathbf{y}^2\} \tag{1.44}$$

where $\mathbf{y}^2$ is determined by (1.39).

Clearly, equation (1.39) satisfies the first self-consistency condition of Section 1.11. However, this alone is insufficient to qualify (1.39) as the input-output-state relationship for $\mathcal{A}(t_0)$. (But it will be seen later that if $\mathcal{A}(t_0)$ admits to the response separation property, highlighted in Chapter 2, then the composite state vector

$$\mathbf{x}(t) = (\mathbf{x}^1(t), \mathbf{x}^2(t), \ldots, \mathbf{x}^N(t))$$

for $\mathcal{A}_1, \mathcal{A}_2, \ldots, \mathcal{A}_N$ qualifies as the state vector of $\mathcal{A}(t_0)$ at time t.)

It is important to note that the composite state vector for the free product $\mathcal{A}_1$ and $\mathcal{A}_2$, i.e.

$$\mathbf{x}(t) = (\mathbf{x}^1(t), \mathbf{x}^2(t))$$

is unlike the initial state vector $(\mathbf{x}_0^1, \mathbf{x}_0^2)$. The initial state vector ranges over the product space $\Psi^1 \times \Psi^2$. The vector $(\mathbf{x}^1(t), \mathbf{x}^2(t))$, which is the state of $\mathcal{A}(t_0)$ at time t, ranges over a subset of $\Psi^1 \times \Psi^2$. This subset $\Psi(t_0,t)$ is

$$\Psi(t_0,t) = \{(\mathbf{x}^1(t), \mathbf{x}^2(t))\} \tag{1.45}$$

where $\mathbf{x}_0^1$ and $\mathbf{x}_0^2$ are elements of Ψ^1 and Ψ^2, respectively, and the states $\mathbf{x}^1(t)$ and $\mathbf{x}^2(t)$ are given by (1.42) and (1.43), respectively. The arguments t_0 and t of the subset serve to show the subset's dependency on both t_0 and t.

Typically, the free product of 3 objects $\mathcal{A}_1$, $\mathcal{A}_2$, $\mathcal{A}_3$ is

$$\mathbf{y}^3 = \mathbf{y}[\mathbf{x}_0^3; \mathbf{y}(\mathbf{x}_0^2; \mathbf{y}(\mathbf{x}_0^1; \mathbf{u}^1))] \tag{1.46}$$

Note that (1.46) is associative. $(\mathcal{A}_3\mathcal{A}_2)\mathcal{A}_1$ and $\mathcal{A}_3(\mathcal{A}_2\mathcal{A}_1)$ both characterize the same input-output-state relation. Obviously, (1.46) can be extended to N objects.

1.15 TANDEM COMBINATIONS—CONSTRAINED

The idea of an initially free tandem combination of objects provides a convenient way of describing the constrained combination, i.e., the combination of, say $\mathcal{A}_1$ and $\mathcal{A}_2$, in which the input to $\mathcal{A}_2$ is constrained for all time t to be the output of $\mathcal{A}_1$. Such a combination, which will be designated as the combination $\mathcal{A}$, is depicted in Figure 1.8 where the switch Q is closed and u^1 is applied to $\mathcal{A}_1$ afterwards.

The constraint $\mathbf{u}^2(t) = \mathbf{y}^1(t)$ for all t is regarded as the limiting form (as $t_0 \to -\infty$) of the constraint $\mathbf{u}^2(t) = \mathbf{y}^1(t)$ for $t > t_0$. The implication here is that the state space of $\mathcal{A}$ is characterized by the limit, if it exists,

$$\Psi(t) = \lim_{t_0 \to -\infty} \Psi(t_0,t) \tag{1.47}$$

The exact meaning of (1.47) and the conditions under which it exists are complicated and unsettled questions. However, for our needs we will interpret (1.47) to mean that for each t we can approach any point in $\Psi(t)$ arbitrarily closely via a point in $\Psi(t_0,t)$. For systems of interest (differential systems) the state at time t is defined by a finite number of derivatives of the input and output at time t. $\Psi(t)$ can be determined through the process of elimination of identical variables from expressions for states of $\mathcal{A}_1$ and $\mathcal{A}_2$.

Based on the above observations the combination $\mathcal{A}$ can be regarded as the limiting form of the free product $\mathcal{A}(t_0)$ as $t_0 \to -\infty$. The state space of $\mathcal{A}$ is a subset Ψ of $\Psi^1 \times \Psi^2$. We have

$$\Psi(t) = \lim_{t_0 \to -\infty} \{(\mathbf{x}^1(t),\mathbf{x}^2(t))\} \tag{1.48}$$

where $\mathbf{x}_0^1$, $\mathbf{x}_0^2$ and $\mathbf{u}^1$ are elements of Ψ^1, Ψ^2 and $R[\mathbf{u}^1]$, respectively. $\mathbf{x}^1(t)$ and $\mathbf{x}^2(t)$ can be determined by expressions (1.42) and (1.43), respectively. The corresponding input-output-state relationship for $\mathcal{A}$ is of form (1.39) and (1.46), with $(\mathbf{x}_0^1,\mathbf{x}_0^2)$ ranging over Ψ.

Clearly, for the objects $\mathcal{A}_1,\mathcal{A}_2,\dots,\mathcal{A}_N$ the above analysis can be expanded as appropriate.

1.16 SYSTEM

We now return to our definition[1] of a system as given in Section 1.2. A system S is a partially interconnected set of objects, called components. These components may be oriented or nonoriented; they may be finite or infinite in number; and each may be associated with a finite or infinite number of terminal variables.

A system may be regarded as a single object. Conversely, any single object may be regarded as a system. If a collection of objects $\mathcal{A}_1, \mathcal{A}_2, \ldots, \mathcal{A}_N$ is considered as a single system the set of terminal variables of the system S is the union of the terminal variables of its components. If the components of S are oriented, then so is S. The aggregate of shared terminal variables and the non-shared output variables of the various components together comprise the output variables of S.

1. *op. cit.*, p. 65.

2
State Equations

2.1 INTRODUCTION

The intent of Chapter 1 was a more careful definition of a system than is found in the texts of physics, mechanics and control theory. The building block of the defintition was the oriented object, i.e., the sets of ordered input-output pairs. The notion of state allows for labeling the pairs, or the parametrization of the space of input-output pairs. The discussions of this chapter will expand on the idea of system state. Specifically, the fundamental properties of the self consistency conditions will be examined; the most important being the state separation property. This property leads directly to an input-output-state relation, in canonical form, for linear differential systems. The definition of state, as formulated in Chapters 1 and 2 is of general applicability. It provides a basis for defining the states of complex systems comprised of both linear and nonlinear components.

2.2 SEPARATION PROPERTY

In Sections 1.7 through 1.11 we defined $\mathbf{x}(t)$, the state of $\mathfrak{A}$ at time t, where $t > t_0$, as an element of the intersecting A sets $A(\mathbf{x}_0;\mathbf{u}^0\mathbf{u}^1)$. These sets are created at fixed t by varying $\mathbf{u}^1$ over the input segment space. By its definition A is the set of all $\mathbf{x}$ satisfying the relation

$$\mathbf{y}(\mathbf{x}_0;\mathbf{u}^0\mathbf{u}^1) = \mathbf{y}(\mathbf{x}_0;\mathbf{u}^0)\mathbf{y}(\mathbf{x};\mathbf{u}^1) \tag{2.1}$$

or

$$\mathbf{y}(\mathbf{x}(t_0);\mathbf{u}^0\mathbf{u}^1) = \mathbf{y}(\mathbf{x}(t_0);\mathbf{u}^0)\mathbf{y}(\mathbf{x}(t);\mathbf{u}^1) \tag{2.2}$$

for all $\mathbf{x}(t_0)$, $\mathbf{u}^0$ and $\mathbf{u}^1$. Relations (2.1) and (2.2) are interpreted to mean that given an input consisting of segment $\mathbf{u}^0$ followed by segment $\mathbf{u}^1$ the response of $\mathcal{Q}$ starting in state $\mathbf{x}(t_0)$ is response segment $\mathbf{y}(\mathbf{x}(t_0);\mathbf{u}^0)$ *followed* by response segment $\mathbf{y}(\mathbf{x}(t);\mathbf{u}^1)$, where $\mathbf{x}(t)$ is the state of $\mathcal{Q}$ at time $t > t_0$. This property of $\mathbf{x}(t)$ is referred to as the *separation property*.

By specifically designating $\mathbf{x}_0^1$ as the terminal state into which $\mathbf{x}_0$ is taken by $\mathbf{u}^0$ equation (2.1) becomes

$$\mathbf{y}(\mathbf{x}_0;\mathbf{u}^0\mathbf{u}^1) = \mathbf{y}(\mathbf{x}_0;\mathbf{u}^0)\mathbf{y}(\mathbf{x}_0^1;\mathbf{u}^1) \tag{2.3}$$

Equations (2.1), (2.2) and (2.3) can be rewritten as an expression for $\mathbf{y}(t)$ rather than the response segment $\mathbf{y}(t_0,t]$. If we let the variable ξ be the intermediate time between t_0 and t_1 we have (in the limit) the identity corresponding to (2.3), i.e.,

$$\mathbf{y}(\mathbf{x}(t_0);\mathbf{u}(t_0,t]) = \mathbf{y}(\mathbf{x}(\xi);\mathbf{u}(\xi,t]) \qquad t_0 < \xi \leqslant t \tag{2.4}$$

The expression $\mathbf{y}(\mathbf{x}(\xi);\mathbf{u}(\xi,t])$ is the response of $\mathcal{Q}$ at time t to $\mathbf{u}(\xi,t]$, with $\mathcal{Q}$ initially in $\mathbf{x}(\xi)$ and where $t_0 < \xi \leqslant t$.

The analyses of (2.1)–(2.4) places in evidence a distinct connection between the self-consistency conditions and the separation property. The relevant facts are summarized as follows. If a relation of the general form

$$\mathbf{y}(t) = \mathbf{y}(\mathbf{x};\mathbf{u}) \tag{2.5}$$

has the response separation property (for all $\mathbf{x}$ and $\mathbf{u}$),

$$\mathbf{y}(\mathbf{x}_0;\mathbf{u}(t_0,t]) = \mathbf{y}(\mathbf{x}_0^1;\mathbf{u}(\xi,t]) \qquad t_0 < \xi \leqslant t \tag{2.6}$$

(where $\mathbf{x}_0^1$ depends only on $\mathbf{x}_0$ and $\mathbf{u}$) then we assert that (2.5) satisfies the three self consistency conditions, with $\mathbf{x}_0$ and $\mathbf{x}_0^1$ being the states of $\mathcal{Q}$ at time t_0 and ξ, respectively. Further, equation (2.5) qualifies as an input-output-state relationship.

Clearly, the first self-consistency condition is satisfied since $\mathbf{y}(t)$ is uniquely determined by $\mathbf{x}_0$ and $\mathbf{u}$. The separation property as stated by (2.6) is, in fact, a statement of the second self-consistency condition (see Section 1.12). By adding the qualification that $\mathbf{x}_0^1$ depends only on $\mathbf{x}_0$ and $\mathbf{u}$, then every set $A(\mathbf{x}_0;\mathbf{u}^0\mathbf{u}^1)$ will contain $\mathbf{x}_0^1$. Therefore the intersection of the A sets formed by varying $\mathbf{u}^1$ over the input segment space will be nonempty. Thus, the third self-consistency condition is satisfied. The importance of the separation property is that for a *given input-output-state relation of form (2.5) verification of the self-consistency conditions is reduced to demonstrating that (2.5) has separation property (2.6).*

The separation property (2.4) can logically be extended to include the *state* of $\mathcal{Q}$. Given that the input-output-state relation is of form (2.5), i.e.,

$$\mathbf{y}(t) = \mathbf{y}(\mathbf{x}(t_0);\mathbf{u}(t_0,t])$$

the state equation induced by (2.5) is

$$\mathbf{x}(t) = \mathbf{x}(\mathbf{x}(t_0);\mathbf{u}(t_0,t]) \tag{2.7}$$

In keeping with (2.4) we assert that state equation (2.7) has the *state separation property*

$$\mathbf{x}(\mathbf{x}(t_0);\mathbf{u}(t_0;t]) \simeq \mathbf{x}(\mathbf{x}(\xi);\mathbf{u}(\xi,t]) \qquad t_0 < \xi \leqslant t \tag{2.8}$$

for all $\mathbf{x}(t_0)$ and ξ. This property of state equation (2.7) is one of its key characteristics.

To illustrate both the response and state separation properties of (2.4) and (2.8) let $\mathcal{Q}$ be characterized by the input-output state relation

$$\begin{aligned} y(t) &= y(x(t_0);u(t_0,t]) \\ &= x(t_0)e^{-(t-t_0)} + \int_{t_0}^{t} e^{-(t-\xi)}u(\xi)d\xi \end{aligned} \tag{2.9}$$

The state equation induced by (2.9) is

$$x(t) = x(t_0)e^{-(t-t_0)} + \int_{t_0}^{t} e^{-(t-\xi)}u(\xi)d\xi \tag{2.10}$$

Specifying the integration variable as τ, where $t_0 \leqslant \tau \leqslant t$, equation (2.9) can be written as

$$y(t) = x(t_0)e^{-(t-t_0)} + \int_{t_0}^{\tau} e^{-(t-\xi)}u(\xi)d\xi + \int_{\tau}^{t} e^{-(t-\xi)}u(\xi)d\xi \quad (2.11)$$

However, from (2.10)

$$x(\tau) = x(t_0)e^{-(\tau-t_0)} + \int_{t_0}^{\tau} e^{-(t-\xi)}u(\xi)d\xi \quad (2.12)$$

Thus, response (2.9) becomes

$$y(t) = x(\tau)e^{-(t-\tau)} + \int_{\tau}^{t} e^{-(t-\xi)}u(\xi)d\xi$$

$$= y(x(\tau);u(\tau,t)) \qquad t_0 < \tau \leqslant t \quad (2.13)$$

which illustrates the response separation property. On returning to (2.10) the state separation property (2.8) assumes the form

$$x(t) = x(\tau)e^{-(t-t_0)} + \int_{t_0}^{\tau} e^{-(t-\xi)}u(\xi)d\xi$$

$$= x(x(\tau);u(\tau,t)) \quad (2.14)$$

2.3 STATE EQUATIONS

The ideas of Sections 2.2 can be combined into the following THEOREM 2.1: If $\mathcal{A}$ is characterized by relations of the form

$$\mathbf{x}(t) = \mathbf{x}(\mathbf{x}(t_0);\mathbf{u}(t_0,t]) \qquad t > t_0 \quad (2.15)$$

$$\mathbf{y}(t) = \mathbf{y}(\mathbf{x}(t);\mathbf{u}(t),t) \quad (2.16)$$

where $\mathbf{x}(t)$ has the state separation property (2.8), then $\mathbf{x}(t)$ can be the state of

$\mathfrak{A}$ at time t, and (2.15) and (2.16) the *state equations* of $\mathfrak{A}$. Proof of this theorem is seen by substituting (2.15) into (2.16) giving

$$\mathbf{y}(t) = \mathbf{y}[\mathbf{x}(\mathbf{x}(t_0);\mathbf{u}(t_0,t]),\mathbf{u}(t),t] \tag{2.17}$$

which is of form (2.5), i.e.,

$$\mathbf{y}(t) = \mathbf{y}(\mathbf{x}(t_0);\mathbf{u}(t_0,t]) \tag{2.18}$$

By virtue of its state separation property equation (2.15) can be rewritten as

$$\mathbf{x}(\mathbf{x}(t_0);\mathbf{u}(t_0,t]) \simeq \mathbf{x}(\mathbf{x}(\xi);\mathbf{u}(\xi,t]) \tag{2.19}$$

On substituting (2.19) into (2.17) we have

$$\mathbf{y}(t) = \mathbf{y}[\mathbf{x}(\mathbf{x}(\xi);\mathbf{u}(\xi,t]),\mathbf{u}(t),t] \tag{2.20}$$

which is of the form

$$\mathbf{y}(t) = \mathbf{y}(\mathbf{x}(\xi);\mathbf{u}(\xi,t]) \tag{2.21}$$

However, (2.21) can also be derived directly from (2.17) by replacing t_0 with ξ. Thus,

$$\mathbf{y}(\mathbf{x}(t_0);\mathbf{u}(t_0,t]) = \mathbf{y}(\mathbf{x}(\xi);\mathbf{u}(\xi,t]) \tag{2.22}$$

and $\mathfrak{A}$ has response separation property (2.4). Further, since $\mathbf{x}(\xi)$ is determined by $\mathbf{x}(t_0)$ and $\mathbf{u}(t_0,t]$, it follows that: (2.18) is an input-output-state relationship for $\mathfrak{A}$, with $\mathbf{x}(t)$ being the state of $\mathfrak{A}$ at time t; equations (2.15) and (2.16) are the state equations.

Theorem 2.1 offers further insight into the relationship between the state separation property and the self-consistency conditions; it provides an effective way of verifying that the self-consistency conditions are satisfied. An immediate inference from the theorem is seen in the following COROLLARY: If $\mathfrak{A}$ is characterized by *differential equations* of *canonical* form

$$\dot{\mathbf{x}}(t) = \mathbf{x}(\mathbf{x}(t),\mathbf{u}(t),t) \tag{2.23}$$

$$\mathbf{y}(t) = \mathbf{y}(\mathbf{x}(t),\mathbf{u}(t),t) \tag{2.24}$$

where (2.23) has a unique solution for $\mathbf{x}(t)$, then $\mathbf{x}(t)$ is the state of $\mathfrak{A}$ at time t.

To prove this corollary we first integrate (2.23) between the limits t_0 and t:

$$\mathbf{x}(t) = \mathbf{x}(t_0) + \int_{t_0}^{t} \mathbf{x}(\mathbf{x}(\xi),\mathbf{u}(\xi),\xi)d\xi \tag{2.25}$$

which is an implicit form of (2.15). (By hypothesis $\mathbf{x}(t)$ is determined by $\mathbf{x}(t_0)$ and $\mathbf{u}(t_0,t]$.) Equation (2.25) has the separation property

$$\mathbf{x}(t) = \mathbf{x}(\tau) + \int_{\tau}^{t} \mathbf{x}(\mathbf{x}(\xi),\mathbf{u}(\xi),\xi)d\xi$$

The conclusion of the proof of this corollary follows from the parent theorem.

State equations (2.23) and (2.24) are quite general and cannot be readily used in their present form. By casting them in a more explicit form, which is applicable only to *linear systems*, we can state that if $\mathcal{C}$ is characterized by *linear differential state equations*

$$\dot{\mathbf{x}}(t) = \mathbf{A}(t)\mathbf{x}(t) + \mathbf{B}(t)\mathbf{u}(t) \tag{2.26}$$

$$\mathbf{y}(t) = \mathbf{C}(t)\mathbf{x}(t) + \mathbf{D}(t)\mathbf{u}(t) \tag{2.27}$$

where $\mathbf{A}(t)\ldots\mathbf{D}(t)$ are time-dependent (but need not be) matrices then $\mathbf{x}(t)$ qualifies as the state of $\mathcal{C}$ at time t. Equations (2.26) and (2.27) are the *canonical state equations* of the linear system $\mathcal{C}$. They are directly deducible[1] from (2.23) and (2.24), respectively.

1. See, for example, *op. cit.* pp. 82-83.

3

Time Invariance, Linearity and Basis Functions

3.1 INTRODUCTION

To further solidify the ideas developed thus far we will establish, by definition, certain "benchmarks" relative to which analytic discussion can be more meaningful. In particular, these benchmarks will include definitions of the *zero*, *ground* and *equilibrium* states.

The *zero state*, say state θ, is defined as that state where, for all t_0, the system response to zero input, starting in state θ, is a zero-valued output (null function). Symbolically, if, for all t_0,

$$\mathbf{y}(t) = \mathbf{y}(\theta;0) = 0 \tag{3.1}$$

where $\mathbf{x}(t_0) = \theta$, then θ is the zero state. The *zero state response* of the system at time t to input $\mathbf{u}$ is the response with the system initially in its zero state θ; i.e.,

$$\begin{aligned} \mathbf{y}(t) &= \mathbf{y}(\theta;\mathbf{u}) \\ &= \mathbf{y}(\mathbf{u}) \end{aligned} \tag{3.2}$$

The idea of the *ground state* coincides with that of the zero state. Consider a system at time t_0 in the initial state $\mathbf{x}(t_0) = \mathbf{x}_0$ and let the applied input be zero-valued, i.e., $\mathbf{u} = 0$. We want to examine $\mathbf{x}(t)$ as $t \to \infty$. There are two possibilities: (1) $\mathbf{x}(t)$ may converge to a fixed state, say $\mathbf{x}_1$, or (2) $\mathbf{x}(t)$ may not converge. We will restrict our attention to (1). In case (1) two possibilities again exist: (a) $\mathbf{x}_1$ depends on $\mathbf{x}_0$, or (b) $\mathbf{x}_1$ is independent of $\mathbf{x}_0$. The ground state of interest is condition 1(b); it is said to be the limiting terminal state into which the system eventually settles when no input is applied. Symbolically, to say state $\mathbf{x}_1$ is the ground state implies that

$$\mathbf{x}_1 = \lim_{t \to \infty} \mathbf{x}(\mathbf{x}_0;0) \tag{3.3}$$

where $\mathbf{x}_1$ is independent of $\mathbf{x}_0$. The limit (3.3) is unique. Accordingly, if $\mathbf{x}_1$ exists it also is unique. In conjunction with (3.2) the *ground state response* of the system has the same meaning as the zero state response, with the exception that the initial state is $\mathbf{x}_0$ instead of θ. At time t the ground state response of the system to input $\mathbf{u}$ is

$$\mathbf{y}(t) = \mathbf{y}(\mathbf{x}_0;\mathbf{u}) \tag{3.4}$$

In addition to the zero and ground states it is meaningful to establish the identity of the *equilibrium* state. Essentially, a state, say $\bar{\theta}$, is called an equilibrium state if it does not change for zero input. Symbolically, to say $\bar{\theta}$ is the equilibrium state is to say that

$$\bar{\theta} = \mathbf{x}(\bar{\theta};0) \qquad t \geqslant t_0 \tag{3.5}$$

where $\mathbf{x}(\bar{\theta};0)$ is the system state at time $t \geqslant t_0$, given that the state at time t_0 is $\bar{\theta}$ when the zero-valued input is applied.

3.2 TIME INVARIANCE

In the sequel the specific systems of interest will be those that are *time-invariant* and *linear*. Time invariance can be conveniently discussed in terms of a time translation operator $\mathbf{T}_\Delta$. The action of the operator on an operand is to shift the operand by a fixed amount of time Δ along the time axis. Positive and negative Δ correspond, respectively, to a delay or increase of Δ units. For $\Delta = 0$, $\mathbf{T}_\Delta$ corresponds to the identity operator $\mathbf{I}$. This action on input $\mathbf{u}$ is depicted in Figure 3.1.

Consider a system in its initial state $\mathbf{x}_0$, but not necessarily the zero state θ.

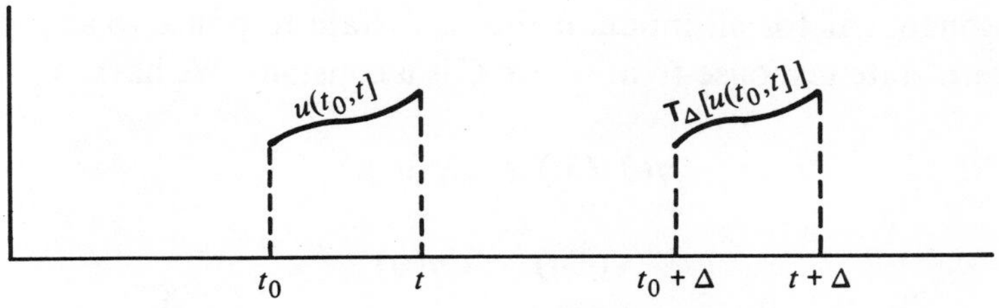

Figure 3.1 Time translate operator $\mathbf{T}_\Delta$.

The system is said to be *time invariant* with respect to the initial state $\mathbf{x}_0$ if, for all starting states $\mathbf{x}_0$, all inputs $\mathbf{u}$ and all time shifts Δ,

$$\mathbf{T}_\Delta[\mathbf{y}(\mathbf{x}_0;\mathbf{u})] = \mathbf{y}(\mathbf{x}_0;\mathbf{T}_\Delta\mathbf{u}) \tag{3.6}$$

Expression (3.6) is interpreted to mean that the response to $\mathbf{T}_\Delta\mathbf{u}$ with the system initially in $\mathbf{x}_0$ is equal to the shifted response $\mathbf{T}_\Delta[\mathbf{y}(\mathbf{x}_0;\mathbf{u})]$ with the system initially in $\mathbf{x}_0$.

As special cases of (3.6) we examine *zero state* and *zero-input time invariance*. A system is said to be *zero state time invariant* if for all inputs $\mathbf{u}$ and time shifts $\pm\Delta$ the zero-state response to $\mathbf{u}$ is a time translate of its zero-state response to the translated input, i.e., for all Δ and $\mathbf{u}$ equation (3.6) becomes

$$\mathbf{y}(\mathbf{T}_\Delta(\theta;\mathbf{u})) = \mathbf{T}_\Delta(\mathbf{y}(\theta;\mathbf{u}))$$

$$\mathbf{y}(\mathbf{T}_\Delta(\mathbf{u})) = \mathbf{T}_\Delta\mathbf{y}(\mathbf{u}) \tag{3.7}$$

Similarly, a system is said to be *zero input time invariant* if for all initial states $\mathbf{x}_0$, all initial times t_0 and all time shifts Δ, the zero input response to $\mathbf{u}$ starting in state $\mathbf{x}_0$ at time $t_0 \pm \Delta$ is identical to the zero input response starting in state $\mathbf{x}_0$ at time t_0. This implies that for all Δ and $\mathbf{x}(t_0)$, equation (3.6) becomes

$$\mathbf{y}(\mathbf{x}(t_0 \pm \Delta);0) = \mathbf{T}_\Delta\mathbf{y}(\mathbf{x}(t_0);0) \tag{3.8}$$

3.3 LINEARITY

The term *linearity* suggests proportionality. Of interest will be the proportionality among system quantities; specifically, their *homogeneous* and *additive* properties. A system is said to be linear if it is both homogeneous and additive.

It is homogeneous if for all inputs $\mathbf{u}$ the zero state response to an input $C\mathbf{u}$ is C times the zero state response to $\mathbf{u}$, where C is a constant. We have

$$\mathbf{y}(\theta; C\mathbf{u}) = C\mathbf{y}(\theta; \mathbf{u})$$

$$\mathbf{y}(C\mathbf{u}) = C\mathbf{y}(\mathbf{u}) \tag{3.9}$$

where $\mathbf{y}(\mathbf{u})$ is the zero state response to $\mathbf{u}$. Similarly, a system is additive if for any pair (or more) of inputs, $\mathbf{u}^1$ and $\mathbf{u}^2$, the zero state response to $\mathbf{u}^1 + \mathbf{u}^2$ is the sum of the zero state responses to $\mathbf{u}^1$ and $\mathbf{u}^2$:

$$\mathbf{y}(\theta; \mathbf{u}^1 + \mathbf{u}^2) = \mathbf{y}(\theta; \mathbf{u}^1) + \mathbf{y}(\theta; \mathbf{u}^2)$$

$$\mathbf{y}(\mathbf{u}^1 + \mathbf{u}^2) = \mathbf{y}(\mathbf{u}^1) + \mathbf{y}(\mathbf{u}^2) \tag{3.10}$$

On combining the ideas of (3.9) and (3.10) a system is said to be *zero state linear* if and only if it is both homogeneous and additive. Therefore,

$$\mathbf{y}(\theta; C(\mathbf{u}^1 + \mathbf{u}^2)) = C\mathbf{y}(\theta; \mathbf{u}^1) + C\mathbf{y}(\theta; \mathbf{u}^2)$$

$$\mathbf{y}(C(\mathbf{u}^1 + \mathbf{u}^2)) = C\mathbf{y}(\mathbf{u}^1) + C\mathbf{y}(\mathbf{u}^2) \tag{3.11}$$

If the initial state is $\mathbf{x}_0$, then the system is *linear with respect to the initial state* $\mathbf{x}_0$ if

$$\mathbf{y}(\theta; C(\mathbf{u}^1 + \mathbf{u}^2)) = C\mathbf{y}(\mathbf{x}_0; \mathbf{u}^1) + C\mathbf{y}(\mathbf{x}_0; \mathbf{u}^2) \tag{3.12}$$

holds for all real constants C and all $\mathbf{u}$ in the input segment space.

As a further consequence of (3.11) and (3.12) it can readily be shown that if a system is zero state linear then it is also linear with respect to all initial states reachable from the zero state. However, it is not linear with respect to all *possible* initial states. To clarify what is meant by linearity we add the property of *zero input linearity*. A system is *zero input linear* if its zero input response is a homogeneous and additive function of the initial states, i.e.,

$$\mathbf{y}(C\mathbf{x}_0; 0) = C\mathbf{y}(\mathbf{x}_0; 0) \tag{3.13}$$

$$\mathbf{y}(\mathbf{x}_0^1 + \mathbf{x}_0^2; 0) = \mathbf{y}(\mathbf{x}_0^1; 0) + \mathbf{y}(\mathbf{x}_0^2; 0) \tag{3.14}$$

for all $\mathbf{x}_0$ (i.e., $\mathbf{x}_0^1, \mathbf{x}_0^2, \ldots$) in the state space Ψ.

Equations (3.13) and (3.14) can be combined giving

$$\mathbf{y}(C(\mathbf{x}_0^1 + \mathbf{x}_0^2); 0) = C\mathbf{y}(\mathbf{x}_0^1; 0) + C\mathbf{y}(\mathbf{x}_0^2; 0) \tag{3.15}$$

Clearly, from (3.12) and (3.15) a general definition of a linear system, which includes both zero state and zero input linearity, can be made as follows: A system is *linear* if and only if

(1) it is linear with respect to all possible initial states, i.e.,

$$\mathbf{y}(\theta; C(\mathbf{u}^1 + \mathbf{u}^2)) = C\mathbf{y}(\mathbf{x}_0; \mathbf{u}^1) + C\mathbf{y}(\mathbf{x}_0; \mathbf{u}^2)$$

(2) it is zero input linear, i.e.,

$$\mathbf{y}(C(\mathbf{x}_0^1 + \mathbf{x}_0^2); 0) = C\mathbf{y}(\mathbf{x}_0^1; 0) + C\mathbf{y}(\mathbf{x}_0^2; 0)$$

As an adjunct to the above definition another basic property for linear systems can be deduced from (3.15). Letting $C = 1$ and $\mathbf{u}^2 = 0$ we have, using (3.2),

$$\mathbf{y}(\mathbf{x}_0; \mathbf{u}) = \mathbf{y}(\mathbf{x}_0; 0) + \mathbf{y}(\theta; \mathbf{u}) \tag{3.16}$$

The system response to $\mathbf{u}$ starting in state $\mathbf{x}_0$ is equal to the zero input response starting in state $\mathbf{x}_0$ plus the zero state response to $\mathbf{u}$. Property (3.16) is known as the *decomposition property*. Since this property was derived from (3.15) it follows then that every linear system has the decomposition property. However, the converse is not necessarily true. We conclude the discussion on linearity by stating that *every linear system must be zero state linear, zero input linear and must possess the decomposition property*.

Equation (3.16) is another form of our initial input-output-state equation (1.6). The importance of linearity is to provide a relative ease in determining the system response to a given $\mathbf{u}$. Through the decomposition property the effect of $\mathbf{u}$ is separated from the initial excitation as represented by $\mathbf{y}(\mathbf{x}_0;0)$. As will be seen later, the zero state response $\mathbf{y}(\theta;\mathbf{u})$ can be reduced to resolving $\mathbf{u}$ into simpler components and determining the zero state response to each component separately.

3.4 ZERO STATE OR IMPULSE RESPONSE

By (3.16) we established that the input-output-state relation for any linear system admits to a representation of the form

$$\mathbf{y}(t) = \mathbf{y}(\mathbf{x}_0; 0) + \mathbf{y}(\theta; \mathbf{u}) \tag{3.16}$$

Focusing our attention on the zero state response, without loss of generality we let $\theta = \mathbf{x}_0$ and $\mathbf{u} = \delta(t - \xi)$, a unit impulse (delta function). We chose

$\mathbf{u} = \delta(t - \xi)$ since any arbitrary signal $\mathbf{u}$ can be reduced to a series of elementary functions, such as an impulse (see Appendices A and B). Therefore the problem of determining the response $\mathbf{y}(\theta;\mathbf{u})$ to an arbitrary input is reduced to finding the system response to the elementary signal. The second term on the right in (3.16) becomes

$$\mathbf{y}(\theta;\mathbf{u}) = \mathbf{y}(\mathbf{u}) = \mathbf{y}(\delta(t - \xi)) \tag{3.17}$$

or in scalar form[1]

$$y(t) = y(\delta(t - \xi)) \tag{3.18}$$

A fixed (time invariant) system can be identified with an operator, say H, wherein the system is entirely characterized by its response at time t to a simple impulse $\delta(t - \xi)$ applied at any instant of time ξ. Accordingly, we can write for (3.18)

$$y(t) = y(u) = H[u(t)] \tag{3.19}$$

$$= H[\delta(t - \xi)] \tag{3.20}$$

$$= h(t, \xi) \tag{3.21}$$

where $h(t,\xi)$ is the zero state response at time t to a unit impulse $\delta(t - \xi)$ applied at time ξ. Accordingly, $h(t,\xi)$ is also called the *impulse response*. From (B.4), however,

$$u(t) = \int_{-\infty}^{\infty} u(\xi)\delta(t - \xi)\,d\xi \tag{3.22}$$

Therefore, substituting (3.22) in (3.19) we have

$$y(t) = H\left[\int_{-\infty}^{\infty} u(\xi)\delta(t - \xi)\,d\xi\right] \tag{3.23}$$

which, due to the homogeneous and additive properties of linear systems, can be written as

1. For convenience in the analysis we will first examine the input-output-state equations in scalar form.

$$y(t) = \int_{-\infty}^{\infty} H[\delta(t-\xi)]u(\xi)d\xi$$

$$= \int_{-\infty}^{\infty} h(t,\xi)u(\xi)d\xi \tag{3.24}$$

Integral (3.24) can be viewed as a summation of responses to impulses applied at time ξ of strength $u(\xi)d\xi$, with ξ varying over the time internal $(-\infty,\infty)$. As a result it is often referred to as the *superposition integral*.

To qualify the zero state response $h(t,\xi)$ as time invariant it must be of the form $h(t-\xi)$. This fact becomes evident by writing (3.24) as

$$y(t) = \int_{t_0}^{t} h(t-\xi)u(\xi)d\xi \tag{3.25}$$

Shifting the input $u(\xi)$ in time by Δ units the output at time $t+\Delta$ becomes

$$y(t) = \int_{t_0+\Delta}^{t+\Delta} h(t+\Delta-\xi)u(\xi-\Delta)d\xi$$

We see that the zero state response at time $t+\Delta$ is equal in magnitude to the zero state response at time t, which satisfies condition (3.7). Hence, we assert that a linear system is zero state time invariant if its impulse response $h(t,\xi)$ is of the form $h(t-\xi)$. Therefore a system obeying the input-output relation (3.19) is time invariant if

$$y(t-\xi) = H[u(t-\xi)] \tag{3.26}$$

For multidimensional inputs, where

$$u(t) = \sum_{i=1}^{n} u_i(t)$$

each component $u_i(t)$ may be expressed as

$$u_i(t) = \int_{-\infty}^{\infty} u_i(\xi)\delta(t-\xi)d\xi \tag{3.27}$$

The response $y_i(t) = y_i(u)$ resulting from input $u_j(t)$ is

$$y_i(t) = \int_{-\infty}^{\infty} h_{ij}(t,\xi)u_j(\xi)d\xi \qquad i = 1,2,\ldots,n \tag{3.28}$$

where $h_{ij}(t,\xi)$ is the response at output terminal i occurring at time t due to a unit impulse applied to input terminal j at time ξ. The resultant output $Y_i(t)$ at the ith terminal due to all inputs $j = 1,2,\ldots,k$, is

$$Y_i(t) = \sum_{j=1}^{k} \int_{-\infty}^{\infty} h_{ij}(t,\xi)u_j(\xi)d\xi \tag{3.29}$$

In matrix from (3.29) becomes

$$\mathbf{y}(t) = \mathbf{y}(\mathbf{u}) = \int_{-\infty}^{\infty} \mathbf{H}(t,\xi)\mathbf{u}(\xi)d\xi \tag{3.30}$$

where $\mathbf{H}$ is the matrix of all zero state responses at time t due to all the unit impulses applied at time ξ.

It is often analytically convenient to express $h(t,\xi)$ in the frequency domain, i.e., as a transfer function. Accordingly, the Laplace transform is the analytic tool. For systems where the impulse response is $h(t-\xi)$ the *transfer function* $H(s)$, where s is the complex frequency, is defined as the Laplace transform of $h(t)$:

$$H(s) = \mathcal{L}\{h(t)\}$$

$$= \int_{-\infty}^{\infty} h(t)e^{-st}dt$$

$$= \int_{-\infty}^{\infty} h(t-\xi)e^{-s(t-\xi)}dt \tag{3.31}$$

3.5 ZERO INPUT RESPONSE AND BASIS FUNCTIONS

We next examine the first term on the right in (3.16), i.e., the zero input response. Reflecting on (3.5) we see that the zero input response is synonomous to the (equilibrium) state at time t beginning in $\mathbf{x}_0$ at time t_0 when the zero-valued input is applied. Our goal will be to more explicitly define $\mathbf{y}(\mathbf{x}_0;0)$ in terms of the basis vectors which span the state space Ψ. In doing so, however, we will revert to scalar quantities to make the analysis more convenient.

Let the linear vector space defining the vector $\mathbf{y}(\mathbf{x}_0;0)$ be spanned by the k basis vectors $\mathbf{a}_i$ ($i = 1,2,\ldots,k$), where the scalar product of any two *unit* vectors $\mathbf{a}_i$ and $\mathbf{a}_j$ obeys the relation

$$\mathbf{a}_i \cdot \mathbf{a}_j = \langle \mathbf{a}_i | \mathbf{a}_j \rangle = \delta_{ij} \tag{3.32}$$

The vector $\mathbf{x}(t)$ can be represented as

$$\mathbf{x}(t) = \sum_{i=1}^{k} \mathbf{a}_i x_i(t) \tag{3.33}$$

The scalar functional of the vector $\mathbf{x}(t)$ can be written as the linear combination

$$F(\mathbf{x}(t)) = \sum_{i=1}^{k} \varphi_i x_i \tag{3.34}$$

where the φ_i are scalar constants or functions. Consequently, the scalar-valued zero input response must be of the form

$$y(\mathbf{x}_0;0) = \sum_{i=1}^{k} \varphi_i(t_0,t) x_i(t_0) \tag{3.35}$$

where

$$\mathbf{x}_0 = (x_0^1, x_0^2, \ldots, x_0^k)$$

and $\varphi_i(t_0,t)$ ($i = 1,2,\ldots,k$), play the role of constants for fixed t and t_0.

The φ_i have a simple interpretation. They form the elements of a k-dimensional vector $\boldsymbol{\phi}$, where

$$\begin{aligned}\boldsymbol{\phi} &= \sum_{i=1}^{k} \mathbf{a}_i \, \varphi_i(t_0,t) \\ &= (\varphi_1(t_0,t), \varphi_2(t_0,t), \ldots, \varphi_k(t_0,t))\end{aligned} \tag{3.36}$$

and correspond to the $\mathbf{a}_i$ of (3.33). Thus, we can write

$$y(\mathbf{a}_i;0) = \varphi_i(t_0,t) \qquad i = 1,2,\ldots,k \tag{3.37}$$

that is, $\varphi_i(t_0,t)$ is the zero input response starting in state $\mathbf{a}_i$ at time t_0.

Since, by definition, the $\mathbf{a}_i$ are linearly independent, we demand that the functions φ_i also be linearly independent. They, therefore, constitute a set of *basis functions* for the system. The vector $\boldsymbol{\phi}$ defined by (3.36) is the *basis function vector* for the system. Accordingly, the scalar-valued zero input response (3.35) can also be written as

$$y(\mathbf{x}_0;0) = \langle \boldsymbol{\phi} | \mathbf{x}_0 \rangle \tag{3.38}$$

3.6 INPUT-OUTPUT-STATE RELATION AND BASIS FUNCTIONS

Returning to (3.16) the scalar form of the input-output-state relation for a linear, time invariant system admits to

$$y(t) = \langle \boldsymbol{\phi} | \mathbf{x}_0 \rangle + \int_{t_0}^{t} h(t - \xi)u(\xi)d\xi \tag{3.39}$$

As such (3.39) satisfies the self consistency conditions of Sections 1.11, 1.12 and 1.13. Equivalently, it has the separation properties of Sections 2.2 and 2.3.

The choice of basis functions φ_i which satisfy (3.39) is constrained, since they must induce a relationship between themselves and the impulse response $h(t - \xi)$. We shall now embark on examining these constraints more closely for the purpose of deducing state equations in a more explicit form.

We begin by assuming that the basis functions φ_i are infinitely differentiable and that the derivatives are continuous over the (finite) observation time of interest. The response separation property of Sections 2.2 and 2.3 specify that in the time interval $t_0 \leqslant \tau \leqslant t$ the *zero-input* response at time t starting with state $\mathbf{x}(t_0)$ must be the same as the response at time t starting in state $\mathbf{x}(\tau)$. Using this property the basis functions must therefore satisfy the (time translational) requirement

$$\langle \boldsymbol{\phi}(t - t_0) | \mathbf{x}(t_0) \rangle = \langle \boldsymbol{\phi}(t - \tau) | \mathbf{x}(\tau) \rangle \tag{3.40}$$

for all t_0, t and τ. Clearly, on letting $t_0 = 0$ and $\mathbf{x}(t_0) = \mathbf{a}_1$, equation (3.40) reduces to (for an n-dimensional vector)

$$\varphi_1 = \sum_{i=1}^{n} \varphi_i(t-\tau)x_i(\tau) \tag{3.41}$$

Equation (3.41) implies that each basis function φ_i can be represented as a linear combination of the set of delayed bases functions $\varphi_i(t-\tau)$, $(i = 1,2,\ldots,n)$. This is the translational property of basis functions. On differentiating (3.40) $n-1$ times we have the following n equations:

$$\langle \boldsymbol{\phi}^{(\alpha-1)}(t-t_0)|\mathbf{x}(t_0)\rangle = \langle \boldsymbol{\phi}^{(\alpha-1)}(t-\tau)|\mathbf{x}(\tau)\rangle \tag{3.42}$$

The superscript signifies the differentiation. Equations (3.42) can be written in compact matrix form as

$$\Phi(t-t_0)\mathbf{x}(t_0) = \Phi(t-\tau)\mathbf{x}(\tau) \tag{3.43}$$

where Φ is an $n \times n$ matrix. The rows of Φ are the vectors $\boldsymbol{\phi},\boldsymbol{\phi}^{(1)},\ldots,\boldsymbol{\phi}^{(n-1)}$:

$$\Phi(t) = \begin{bmatrix} \boldsymbol{\phi} \\ \boldsymbol{\phi}^{(1)} \\ \vdots \\ \boldsymbol{\phi}^{(n-1)} \end{bmatrix} = \begin{bmatrix} \varphi_1(t) & \varphi_2(t) & \cdots & \varphi_n(t) \\ \varphi_1^{(1)}(t) & \varphi_2^{(1)}(t) & \cdots & \varphi_n^{(1)}(t) \\ \cdots & \cdots & \cdots & \cdots \\ \varphi_1^{(n-1)}(t) & \varphi_2^{(n-1)}(t) & \cdots & \varphi_n^{(n-1)}(t) \end{bmatrix} \tag{3.44}$$

By setting $t = \tau$ in the right-hand side of (3.43) the relationship between $\mathbf{x}(t)$ and $\mathbf{x}(t_0)$ is established as

$$\Phi(0)\mathbf{x}(t) = \Phi(t-t_0)\mathbf{x}(t_0) \tag{3.45}$$

If $\Phi(0)$ is nonsingular then $\mathbf{x}(t)$ is determined uniquely. The basis functions $\varphi_1,\varphi_2,\ldots,\varphi_n$ can be normalized giving $\Phi(0) = \mathbf{I}$, where $\mathbf{I}$ is the identity matrix. The normalized set of basis functions φ_i' is found through the transformation

$$\boldsymbol{\phi}' = \boldsymbol{\phi}\Phi^{-1}(0) \tag{3.46}$$

Equation (3.45) thereby reduces to

$$\mathbf{x}(t) = \Phi(t-t_0)\mathbf{x}(t_0) \tag{3.47}$$

We see from (3.47) that $\Phi(t - t_0)$ is an operator which transforms the initial state $\mathbf{x}(t_0)$ into the state $\mathbf{x}(t)$. Hence, $\Phi(t - t_0)$ is referred to as the *state transition matrix*.

3.7 STATE EQUATIONS

In the preceding discussion the separation property of the zero input response was used to establish the basis function (derivatives) composition of the state transition operator Φ. We will now consider the more general situation of a non-zero input and the relations between the state transition matrix (basis function) and the impulse response $h(t,\xi)$.

In the general case the response separation property of Sections 2.2 and 2.3 imply that the input-output-state equation (3.39) must satisfy the time translation identity

$$\langle \phi(t - t_0) | \mathbf{x}(t_0) \rangle + \int_{t_0}^{t} h(t - \xi) u(\xi) d\xi$$

$$= \langle \phi(t - \tau) | \mathbf{x}(\tau) \rangle + \int_{\tau}^{t} h(t - \xi) u(\xi) d\xi \qquad (3.48)$$

for all t_0, all $t \geqslant t_0$ and all τ, where $t_0 \leqslant \tau \leqslant t$. (Note that by letting $u(t) = \delta(t)$ and setting $\mathbf{x}(t_0) = 0$, $t_0 = 0$, equation (3.48) yields

$$h(t) = \langle \phi(t - \tau) | \mathbf{x}(\tau) \rangle \qquad (3.49)$$

This shows that the impulse response is a linear combination of the basis functions.) Differentiating (3.48) with respect to τ gives

$$\frac{d}{d\tau} \int_{t_0}^{t} h(t - \xi) u(\xi) d\xi = \frac{d}{d\tau} \left[\langle \phi(t - \tau) | \mathbf{x}(\tau) \rangle + \int_{\tau}^{t} h(t - \xi) u(\xi) d\xi \right]$$

$$h(t - \tau) u(\tau) = \frac{d}{d\tau} \langle \phi(t - \tau) | \mathbf{x}(\tau) \rangle$$

$$= -\langle \dot{\phi}(t - \tau) | \mathbf{x}(\tau) \rangle + \langle \phi(t - \tau) | \dot{\mathbf{x}}(\tau) \rangle \qquad (3.50)$$

Now differentiating (3.50) $n - 1$ times with respect to t results in n expressions as follows:

$$h^{(\alpha-1)}(t-\tau)u(\tau) = \frac{d}{d\tau}\langle \phi^{(\alpha-1)}(t-\tau)|\mathbf{x}(\tau)\rangle$$

$$= -\langle \dot{\phi}^{(\alpha-1)}(t-\tau)|\mathbf{x}(\tau)\rangle + \langle \phi^{(\alpha-1)}(t-\tau)|\dot{\mathbf{x}}(\tau)\rangle \quad (3.51)$$

where $\alpha = 1,2,\ldots,n$, denotes the order of the time derivatives. The set of equation (3.51) can be represented more compactly in matrix form as

$$\mathbf{h}(t-\tau)u(\tau) = \frac{d}{dt}[\mathbf{\Phi}(t-\tau)\mathbf{x}(\tau)] \quad (3.52)$$

where $\mathbf{\Phi}$ is the state transition matrix defined in (3.44) and $\mathbf{h}$ is the column vector

$$\mathbf{h}(t) = \begin{bmatrix} h(t) \\ h^{(1)}(t) \\ \vdots \\ h^{(\alpha-1)}(t) \end{bmatrix} \quad (3.53)$$

Integrating (3.52) between the limits t_0 and t gives

$$\int_{t_0}^{t} \mathbf{h}(t-\tau)u(\tau)d\tau = \int_{t_0}^{t} d[\mathbf{\Phi}(t-\tau)\mathbf{x}(\tau)$$

$$= \mathbf{\Phi}(t-\tau)\mathbf{x}(\tau)\Big|_{t_0}^{t}$$

$$= \mathbf{\Phi}(0)\mathbf{x}(t) - \mathbf{\Phi}(t-t_0)\mathbf{x}(t_0)$$

Making use of the fact that $\mathbf{\Phi}(0) = \mathbf{I}$ the above integration yields the state equation

$$\mathbf{x}(t) = \mathbf{\Phi}(t-t_0)\mathbf{x}(t_0) + \int_{t_0}^{t} \mathbf{h}(t-\xi)u(\xi)d\xi \quad (3.54)$$

Equation (3.54) is the *state equation* in *explicit form* for a linear time-invariant system. It expresses the system state at time t as a function of the initial state at time t_0 and the input over the interval $[t_0,t]$. For zero input (3.54) reduces to (3.47).

Earlier we established (3.39) as the scalar-valued input-output-state relationship for a linear system. On differentiating (3.39) $n - 1$ times with respect to t we have

$$y(t) = \langle \phi(t - t_0) | \mathbf{x}(t_0) \rangle + \int_{t_0}^{t} h(t - \xi) u(\xi) d\xi$$

$$\cdots\cdots\cdots\cdots\cdots\cdots\cdots\cdots\cdots\cdots\cdots$$

$$y^{(n-1)}(t) = \langle \phi^{(n-1)}(t - t_0) | \mathbf{x}(t_0) \rangle + \int_{t_0}^{t} h^{(n-1)}(t - \xi) u(\xi) d\xi$$

Comparing the results with (3.54) where

$$\mathbf{x}(t) = \Phi(t - t_0)\mathbf{x}(t_0) + \int_{t_0}^{t} \mathbf{h}(t - \xi) u(\xi) d\xi$$

it can be concluded that

$$\mathbf{x}(t) = (y(t), y^{(1)}(t), \ldots, y^{(n-1)}(t)) \qquad (3.55)$$

The above relationship is valid *only* if $h^{(n-1)}(t)$ does not contain delta functions at $t = 0$. Further, this relationship applies only to the system characterized by a differential equation of the form

$$\ell_n y^{(n)} + \ell_{n-1} y^{(n-1)} + \ldots + \ell_0 y = m_k u^{(k)} + m_{k-1} u^{(k-1)} + \ldots + m_0 u$$

where ℓ and m are constant coefficients.

Returning to (3.54) we examine the response when the input is $u = \delta(t)$. Setting $\mathbf{x}(t_0) = 0$ and $t_0 = 0$ we have

$$\mathbf{h}(t) = \mathbf{x}(t) \qquad t \geqslant 0$$

$$\mathbf{h}(t) = 0 \qquad t < 0 \qquad (3.56)$$

Equations (3.56) are interpreted to mean that $\mathbf{h}(t)$ is the system state at time t given that the system was initially in its zero state at time t_0 when excited by the impulse $\delta(t)$. Thus $\mathbf{h}$ is referred to as the *state impulse response*.

State equation (3.54), which expresses the system state at time t in terms of the initial state at time t_0 and the applied input, can be readily reformatted into a differential form. Returning to (3.51) and setting $\tau = t$, we have, remembering that $\mathbf{\Phi}(0) = \mathbf{I}$ if $\mathbf{\Phi}(0)$ is nonsingular,

$$\dot{\mathbf{x}}(t) = \dot{\mathbf{\Phi}}(0)\mathbf{x}(t) + \mathbf{h}(0)u(t) \tag{3.57}$$

which is the state equation in differential form. $\dot{\mathbf{\Phi}}(0)$ relates to the state transition matrix $\mathbf{\Phi}(t)$, where specifically $\mathbf{\Phi}(t)$ is the solution of the differential equation

$$\dot{\mathbf{\Phi}}(t) = \dot{\mathbf{\Phi}}(0)\mathbf{\Phi}(t) \tag{3.58}$$

Thus, for a system characterized by input-output-state relation

$$y(t) = \langle \boldsymbol{\phi}(t - t_0) | \mathbf{x}(t_0) \rangle + \int_{t_0}^{t} h(t - \xi)u(\xi)d\xi \qquad t \geqslant t_0$$

the state equations in differential form are, for all t,

$$\dot{\mathbf{x}}(t) = \dot{\mathbf{\Phi}}(0)\mathbf{x}(t) + \mathbf{h}(0)u(t) \tag{3.59}$$

$$y(t) = \langle \boldsymbol{\phi}(0) | \mathbf{x}(t) \rangle \tag{3.60}$$

Accordingly, the state at time t is, by (3.54),

$$\mathbf{x}(t) = \mathbf{\Phi}(t - t_0)\mathbf{x}(t_0) + \int_{t_0}^{t} \mathbf{h}(t - \xi)u(\xi)d\xi$$

4
Canonical Formulation

4.1 INTRODUCTION

The canonical forms (2.26) and (2.27) of the state equations provide for the central idea of associating a state vector with a linear differential system. The resulting products of the method described below are expressions for the state vector along with expressions for the matrices **A**, **B**, **C** and **D** of (2.26) and (2.27).

In general a linear differential system can be characterized by an input-output relationship of the form

$$L(D)y = M(D)u \tag{4.1}$$

where the operators $L(D)$ and $M(D)$ are

$$L(D) = \ell_n D^n + \ell_{n-1} D^{n-1} + \dots + \ell_0$$

$$M(D) = m_k D^k + m_{k-1} D^{k-1} + \dots + m_0$$

$$D^i = d^i/dt^i \qquad (i = 1,2,\ldots,k,\ldots,n)$$

$$D^{-1} = \int (\;)dt$$

The coefficients ℓ and m are constants, and not necessarily real.[1] Equation (4.1) can also be formulated as

$$y = M(D)u \tag{4.2}$$

for *differential operator* systems, or as

$$L(D)y = u \tag{4.3}$$

for *reciprocal differential operator* systems. Specifically, our procedure of attaching a state vector to differential systems will be to solve expressions (4.1), (4.2) and (4.3), and relate the resulting coefficients ℓ and m to the components of the state vector.

4.2 RECIPROCAL DIFFERENTIAL SYSTEM

For reciprocal differential systems we seek the solution to (4.3):

$$L(D)y = u$$

where

$$L(D) = \ell_n D^n + \ell_{n-1} D^{n-1} + \ldots + \ell_0$$

To solve (4.3) we will make direct use of the Laplace transform. The transform of the nth derivative of a time function $y(t)$ can be written as

$$\mathcal{L}\left\{\frac{d^n y(t)}{dt^n}\right\} = s^n Y(s) - s^{n-1}y(0) - s^{n-2}Dy(0) - \ldots - D^{n-1}y(0)$$

$$= s^n Y(s) - s^{n-1}y(0) - s^{n-2}y^{(1)}(0) - \ldots - y^{(n-1)}(0)$$

1. In general the index used for both coefficients ℓ and m, i.e., ℓ_m and m_k are not the same. It is recognized that the order of the differential operators associated with u and y will be different from one another; however, there is no loss in generality in using a single subscript m. This will be done in the sequel for convenience only.

where s is the complex frequency, $y^{(i)}$ is the ith derivative of $y(0)$, and $Y(s)$ is the transform of $y(t)$. The transform of both sides of (4.3) gives

$$\begin{aligned}(\ell_n s^n + \ell_{n-1}s^{n-1} + \ldots + \ell_0)Y(s) &= U(s) + \ell_n y^{(n-1)}(0)\\ &\quad + (\ell_n s^n + \ell_{n-1}s^{n-1} + \ldots + \ell_0)y^{(n-2)}(0)\\ &\quad + \ldots + (\ell_n s^n + \ell_{n-1}s^{n-1} + \ldots + \ell_0)\end{aligned} \tag{4.4}$$

On letting

$$L(s) = \ell_n s^n + \ell_{n-1}s^{n-1} + \ldots + \ell_0$$

equation (4.4) reduces to

$$Y(s) = \frac{U(s)}{L(s)} + \sum_{i=1}^{n} \frac{\ell_n s^{n-i} + \ldots + \ell_i}{L(s)} y^{(i-1)}(0) \tag{4.5}$$

The inverse Laplace transform of both sides of (4.5) gives the time dependent output as

$$y(t) = \sum_{i=1}^{n} y^{(i-1)}(0)\varphi_i(t) + \int_0^t h(t-\xi)u(\xi)d\xi \tag{4.6}$$

where

$$h(t) = \mathcal{L}\{H(s)\}$$

$$H(s) = 1/L(s)$$

and

$$\varphi_i(t) = \mathcal{L}^{-1}\left\{\frac{\ell_n s^{n-i} + \ell_{n-1}s^{n-1-i} + \ldots + \ell_i}{L(s)}\right\} \tag{4.7}$$

The time functions φ_i satisfy the differential equation

$$L(D)\varphi_i = 0$$

whereas h satisfies

$$L(D)h = 0$$

Clearly the φ_i are linearly independent and qualify as a set of basis functions for the system characterized by (4.3). It follows that (starting at $t_0 = 0$) the input-output pair (u,y) satisfying (4.3) also satisfies (4.6), since (4.6) is an extension of (4.3). A closer examination of this relationship reveals that for every input-output pair satisfying (4.3) there corresponds an n-tuple of complex numbers $(\lambda_1,\lambda_2,\ldots,\lambda_n)$ such that (u,y) satisfies (4.6) for $t \geqslant 0$ with

$$y^{(i-1)}(0) = \lambda_i \qquad i = 1,2,\ldots,n \tag{4.8}$$

(Conversely, for every n-tuple $(\lambda_1,\ldots,\lambda_n)$ the expressions for y given by (4.6), where each $y^{(i-1)}(0)$ is replaced by λ_i ($i = 1,2,\ldots,n$), defines an input-output pair for (4.3); i.e., (4.6) is a solution of (4.3) for each $\lambda_1,\lambda_2,\ldots,\lambda_n$.) By virtue of the constancy of the coefficients of (4.3) and the time translation properties of linear time-invariant systems the time-shifted pair $T_\Delta(u,y)$ satisfying (4.3) also satisfies (4.6). Therefore, the general solution to (4.3) for arbitrary t_0 can be written as

$$y(t) = \sum_{i=1}^{n} y^{(i-1)}(t_0)\varphi_i(t - t_0) + \int_{t_0}^{t} h(t - \xi)u(\xi)d\xi \tag{4.9}$$

where $t \geqslant t_0$.

The state vector $\mathbf{x}(t)$ must now be fixed to the system. This can be done for linear time-invariant systems by relating the constants of (4.3) with the components of $\mathbf{x}(t_0)$. A straightforward way to do this, though not necessarily the most advantageous method for all cases, is to relate the basis functions with the respective φ_i of (4.9). From (3.55) the components of $\mathbf{x}(t_0)$, which are the coefficients of the basis functions, are

$$x_1(t_0) = y(t_0)$$

$$x_2(t_0) = y^{(1)}(t_0)$$

$$\cdots\cdots\cdots\cdots\cdots\cdots$$

$$x_n(t_0) = y^{(n-1)}(t_0)$$

We readily identify $\mathbf{x}(t_0)$ as

$$\mathbf{x}(t_0) = (y(t_0), y^{(1)}(t_0), \ldots, y^{(n-1)}(t_0)) \tag{4.10}$$

Digressing temporarily an alternative method of establishing the components of the state vector $\mathbf{x}(t_0)$ is seen from expression (4.5). The Laplace transform of the respective basis functions gives

$$\frac{s^{n-1}}{L(s)}, \frac{s^{n-2}}{L(s)}, \ldots, \frac{s}{L(s)}, \frac{1}{L(s)}$$

Thus, the components of $\mathbf{x}(t_0)$ are

$$x_1(t_0) = \ell_n y(t_0)$$

$$x_2(t_0) = \ell_n y^{(1)}(t_0) + \ell_{n-1} y(t_0)$$

$$x_3(t_0) = \ell_n y^{(2)}(t_0) + \ell_{n-1} y^{(1)}(t_0) + \ell_{n-2} y(t_0)$$

$$\cdots\cdots\cdots\cdots\cdots\cdots\cdots\cdots\cdots\cdots\cdots\cdots\cdots\cdots$$

$$x_n(t_0) = \ell_n y^{(n-1)}(t_0) + \ell_{n-1} y^{(n-2)}(t_0) + \ldots + \ell_2 y^{(1)}(t_0) + \ell_1 y(t_0)$$

Returning to (4.10) the terms $y^{(i-1)}$ $(i = 1,2,\ldots,n)$ in (4.9) are each replaced by their corresponding expressions from (4.10). Accordingly, the solution to (4.3) for arbitrary t_0 becomes

$$y(t_0) \doteq \langle \phi(t - t_0) | \mathbf{x}(t_0) \rangle + \int_{t_0}^{t} h(t - \xi) u(\xi) d\xi \tag{4.11}$$

which is in agreement with (3.39). By (4.11) the state vector $\mathbf{x}(t_0)$ has been fixed to the system.

To complete the analysis of the reciprocal differential operator system it now remains to show that $\mathbf{x}(t)$ qualifies as a state vector. We form $\dot{\mathbf{x}}(t)$ and $\mathbf{y}(t)$. If the resultant expressions are of the canonical form (2.26) and (2.27),

$$\dot{\mathbf{x}}(t) = \mathbf{A}\mathbf{x} + \mathbf{B}\mathbf{u}$$

$$\mathbf{y}(t) = \mathbf{C}\mathbf{x} + \mathbf{D}\mathbf{u}$$

then $\mathbf{x}(t)$ qualifies as a state vector. Considering only fixed systems (systems where $\mathbf{A}$, $\mathbf{B}$, $\mathbf{C}$ and $\mathbf{D}$ are constant matrices) it follows from (4.10) that at time t

$$\mathbf{x}(t) = (y(t), y^{(1)}(t), \ldots, y^{(n-1)}(t)) \tag{4.12}$$

and

$$\dot{\mathbf{x}}(t) = (y^{(1)}(t), y^{(2)}(t), \ldots, y^{(n)}(t)) \tag{4.13}$$

From (4.3) the general expression involving the nth derivative of $y(t)$ is

$$y^{(n)}(t) = \frac{1}{\ell_n}(u(t) - \ell_{n-1}y^{(n-1)}(t) - \ell_{n-2}y^{(n-2)}(t) - \ldots - \ell_0 y(t))$$

Substituting the above expression into (4.13) gives

$$\dot{\mathbf{x}}(t) = \left(y^{(1)}(t), y^{(2)}(t), \ldots, y^{(n-1)}(t), \frac{1}{\ell_n}[u(t) - \ell_{n-1}y^{(n-1)}(t) - \ldots - \ell_0 y(t)]\right) \tag{4.14}$$

Thus, the components of $\dot{\mathbf{x}}(t)$ are a linear combination of $\mathbf{x}(t)$ and $\mathbf{u}(t)$. Hence, from equations (4.12), (4.13) and (4.14) the components of $\dot{\mathbf{x}}(t)$ are

$$\begin{aligned} \dot{x}_1(t) &= x_2(t) \\ \dot{x}_2(t) &= x_3(t) \\ &\cdots\cdots\cdots\cdots \\ \dot{x}_n(t) &= -\frac{\ell_0}{\ell_n}x_1(t) - \ldots - \frac{\ell_{n-1}}{\ell_n}x_n(t) + \frac{1}{\ell_n}u(t) \\ y(t) &= x_1(t) \end{aligned} \tag{4.15}$$

Equations (4.15) expressed in matrix form become

$$\begin{bmatrix} \dot{x}_1(t) \\ \dot{x}_2(t) \\ \cdots \\ \dot{x}_n(t) \end{bmatrix} = \begin{bmatrix} 0 & 1 & 0 & \ldots & 0 \\ 0 & 0 & 1 & \ldots & 0 \\ \cdots & \cdots & \cdots & \cdots & \cdots \\ -\frac{\ell_0}{\ell_n} & & & & -\frac{\ell_{n-1}}{\ell_n} \end{bmatrix} \begin{bmatrix} x_1(t) \\ x_2(t) \\ \cdots \\ x_n(t) \end{bmatrix} + \begin{bmatrix} 0 \\ 0 \\ \cdots \\ \frac{1}{\ell_n} \end{bmatrix} u(t) \tag{4.16}$$

$$y(t) = [1 \quad 0 \quad \ldots \quad 0] \begin{bmatrix} x_1(t) \\ x_2(t) \\ \cdots \\ x_n(t) \end{bmatrix} \tag{4.17}$$

or

$$\dot{\mathbf{x}}(t) = \mathbf{Ax} + \mathbf{Bu}$$

$$\mathbf{y}(t) = \mathbf{Cx} + \mathbf{Du}$$

$$= \langle \mathbf{C} | \mathbf{x} \rangle \tag{4.18}$$

where

$$\mathbf{A} = \begin{bmatrix} 0 & 1 & 0 & \dots & 0 \\ 0 & 0 & 1 & \dots & 0 \\ \dots & \dots & \dots & \dots & \dots \\ -\dfrac{\ell_0}{\ell} & & & & -\dfrac{\ell_{-1}}{\ell} \end{bmatrix} \qquad \mathbf{B} = \begin{bmatrix} 0 \\ 0 \\ \dots \\ \dfrac{1}{\ell} \end{bmatrix}$$

$$\mathbf{C} = [1 \quad \dots \quad 0] \qquad \mathbf{D} = 0 \tag{4.19}$$

Thus, $\mathbf{x}(t)$ as defined by (4.12) qualifies as the state vector for the reciprocal differential operator system characterized by equation (4.3).

4.3 DIFFERENTIAL OPERATOR SYSTEM

The differential operator system characterized by (4.2) is of interest for two reasons: It is possible to realize any differential system as a combination of both the reciprocal differential and the differential operator types. Secondly, each is the inverse of the other. For the system of (4.2) we seek the general solution to

$$y = M(D)u$$

where

$$M(D) = m_n D^n + m_{n-1} D^{n-1} + \dots + m_0$$

The Laplace transform of both sides of (4.2) gives

$$Y(s) = M(s)U(s) - \sum_{i=1}^{n} (m_n s^{n-i} + \dots + n_i) u^{(i-1)}(0) \tag{4.20}$$

where

$$M(s) = m_n s^n + m_{n-1} s^{n-1} + \dots + m_0 \tag{4.21}$$

and
$$U(s) = \mathcal{L}\{u(t)\} \tag{4.22}$$

The inverse transform of (4.20) gives the time-dependent output, for $t_0 = 0$, as

$$y(t) = \sum_{i=1}^{n} \varphi_i(t)u^{(i-1)}(0) + \int_0^t h(t-\xi)u(\xi)d\xi \tag{4.23}$$

where

$$\varphi_i(t) = \mathcal{L}^{-1}\{m_n s^n + m_{n-1}s^{n-1} + \ldots + n_i\} \tag{4.24}$$

and

$$h(t) = \mathcal{L}^{-1}\{M(s)\} \tag{4.25}$$

From (4.23) the n basis functions φ_i are linearly independent. They represent the zero-input responses to the differential operator system. Similarly, $h(t)$ is the system impulse response. Relating the basis functions with the respective φ_i we let, from (4.23),

$$\begin{aligned} x_1(t) &= u(t) \\ x_2(t) &= u^{(1)}(t) \\ &\cdots\cdots\cdots\cdots \\ x_n(t) &= u^{(n-1)}(t) \end{aligned} \tag{4.26}$$

Thus, the state vector for the system of (4.2) is

$$\mathbf{x}(t) = (u(t), u^{(1)}(t), \ldots, u^{(n-1)}(t)) \tag{4.27}$$

Forming the expression for $\dot{\mathbf{x}}(t)$

$$\dot{\mathbf{x}}(t) = (u^{(1)}(t), u^{(2)}(t), \ldots, u^{(n)}(t))$$

and comparing it with that for $\mathbf{x}(t)$ we see that

$$\begin{aligned} x_1(t) &= u \\ x_2(t) &= u^{(1)} = \dot{x}_1 \\ x_3(t) &= u^{(2)} = \dot{x}_2 \\ &\cdots\cdots\cdots\cdots\cdots \\ x_n(t) &= u^{(n-1)} = \dot{x}_{n-1} \end{aligned} \tag{4.28}$$

Hence, the input output relation (4.2) can be written as

$$\begin{aligned} y &= m_0 u + m_1 u^{(1)} + \ldots + m_n u^{(n)} \\ &= m_0 x_1 + m_1 x_2 + \ldots + m_n \dot{x}_n \\ &= m_0 x_1 + m_1 \dot{x}_1 + \ldots + m_n \dot{x}_n \end{aligned} \tag{4.29}$$

Equations (4.28) and (4.29) comprise the state equations for the reciprocal differential operator system. Equation (4.28) is the corresponding state vector.

In the representation outlined above the state equations representing the differential operator system are *not* of the canonical form

$$\dot{\mathbf{x}} = \mathbf{Ax} + \mathbf{Bu}$$

$$\mathbf{y} = \mathbf{Cx} + \mathbf{Du}$$

The corresponding input-output relation for arbitrary t_0 is obtained by substituting for each $u^{(i-1)}(0)$ in (4.23) the corresponding $x_i(0)$ as determined by (4.28). The result is

$$\begin{aligned} y(t) &= \varphi_1(t - t_0)x_1(t_0) + \ldots + \varphi_n(t - t_0)x_n(t_0) + \int_{t_0}^{t} h(t - \xi)u(\xi)d\xi \\ &= \langle \boldsymbol{\phi}(t - t_0)|\mathbf{x}(t_0)\rangle + \int_{t_0}^{t} h(t - \xi)u(\xi)d\xi \end{aligned} \tag{4.30}$$

4.4 GENERAL SOLUTION

We seek the solution to the generalized equation (4.1):

$$L(D)y = M(D)u$$

where[1]

$$L(D) = \ell_n D^n + \ell_{n-1}D^{n-1} + \ldots + \ell_0$$

$$M(D) = m_n D^n + \ell_{n-1}D^{n-1} + \ldots + \ell_0$$

1. In general the order of the coefficients for ℓ is not the same as those for m. The common order n is being used for convenience only.

The Laplace transform of both sides of (4.1) gives

$$L(s)Y(s) = M(s)U(s) + J(s)$$

$$Y(s) = H(s)U(s) + L^{-1}(s)J(s) \qquad (4.31)$$

The ratio $H(s)$ is defined as

$$H(s) = M(s)/L(s)$$

and the polynomials $L(s)$ and $M(s)$ were previously given as

$$L(s) = \ell_n s^n + \ell_{n-1}s^{n-1} + \ldots + \ell_0$$

$$M(s) = m_n s^n + m_{n-1}s^{n-1} + \ldots + m_0$$

$J(s)$ is the residual polynomial resulting from the Laplace transform of $u(t)$ and $y(t)$ at time $t = 0$. When arranged in powers of s the polynomial $J(s)$ is

$$\begin{aligned} J(s) = {} & s^{n-1}[\ell_n y(0) - m_n u(0)] \\ & + s^{n-2}[\ell_n Dy(0) - m_n Du(0) + \ell_{n-1}y(0) - m_{n-1}u(0)] \\ & \cdots\cdots\cdots\cdots\cdots\cdots\cdots\cdots \\ & + [\ell_n D^{n-1}y(0) - m_n D^{n-1}u(0) + \ldots + \ell_1 y(0) - m_1 u(0)] \end{aligned} \qquad (4.32)$$

The state vector $\mathbf{x}(t)$ can be established by relating the components of $\mathbf{x}(t_0)$ to the components of $J(s)$ for $t_0 = 0$. Equating $x_1(0)$ to the coefficients of s^{n-1}, $x_2(0)$ to the coefficients of s^{n-2}, etc., we have

$$\begin{aligned} x_1 &= \ell y - m_n u \\ x_2 &= \ell_n Dy - m_n Du + \ell_{n-1}y - m_{n-1}u \\ & \cdots\cdots\cdots\cdots\cdots\cdots\cdots\cdots \\ x_n &= \ell_n D^{n-1}y - m_n D^{n-1}u + \ldots + \ell_1 y - m_1 u \end{aligned} \qquad (4.33)$$

Equations (4.33) can also be written as

$$\begin{aligned} x_1 &= \ell_n y - m_n u \\ x_2 &= \dot{x}_1 + \ell_{n-1}y - m_{n-1}u \\ & \cdots\cdots\cdots\cdots\cdots\cdots \\ x_n &= \dot{x}_{n-1} + \ell_1 y - m_1 u \end{aligned} \qquad (4.34)$$

and from (4.1) it is clear that

$$\dot{x}_n = -(\ell_0 y - m_0 u) \tag{4.35}$$

Solving equations (4.34) and (4.35) for the various $\dot{x}_i$ and y in terms of the corresponding x_i and u gives

$$\dot{x}_1 = \frac{1}{\ell_n}[-\ell_{n-1}x_1 + \ell_n x_2 + (\ell_n m_{n-1} - \ell_{n-1}m_n)u]$$

$$\dot{x}_2 = \frac{1}{\ell_n}[-\ell_{n-2}x_1 + \ell_n x_3 + (\ell_n m_{n-2} - \ell_{n-2}m_n)u]$$

$$\cdots\cdots\cdots\cdots\cdots\cdots\cdots\cdots\cdots$$

$$\dot{x}_{n-1} = \frac{1}{\ell_n}[-\ell_1 x_1 + \ell_n x_n + (\ell_n m_1 - \ell_1 m_n)u]$$

$$\dot{x}_n = \frac{1}{\ell_n}[-\ell_0 x_1 + (\ell_n m_0 - \ell_0 m_n)u] \tag{4.36}$$

and

$$y = \frac{1}{\ell_n}[x_1 + m_n u] \tag{4.37}$$

To qualify $\mathbf{x}(t)$ as the state vector it is necessary to demonstrate that $\dot{\mathbf{x}}$, $\mathbf{x}$, y and u satisfy the state equations (2.26) and (2.27). By inspection (4.36) and (4.37) are of the form

$$\dot{\mathbf{x}} = \mathbf{Ax} + \mathbf{Bu}$$

$$\mathbf{y} = \mathbf{Cx} + \mathbf{Du}$$

where

$$\mathbf{A} = \frac{1}{\ell_n}\begin{bmatrix} -\ell_n & \ell_n & 0 & \cdots & 0 \\ -\ell_{n-2} & 0 & \ell_n & \cdots & 0 \\ \cdots & \cdots & \cdots & \cdots & \cdots \\ -\ell_1 & 0 & 0 & \cdots & \ell_n \\ -\ell_0 & 0 & 0 & \cdots & 0 \end{bmatrix} \qquad \mathbf{B} = \frac{1}{\ell_n}\begin{bmatrix} \ell_n m_{n-1} - \ell_{n-1}m_n \\ \ell_n m_{n-2} - \ell_{n-2}m_n \\ \cdots\cdots \\ \cdots\cdots \\ \ell_n m_0 - \ell_0 m_n \end{bmatrix}$$

$$\mathbf{C} = \begin{bmatrix} \frac{1}{\ell_n} & 0 & \dots & 0 \end{bmatrix} \qquad \mathbf{D} = \begin{bmatrix} \frac{m_n}{\ell_n} \end{bmatrix} \tag{4.38}$$

Thus, $\mathbf{x}(t)$ as defined by (4.33) qualifies as the state vector representing a system characterized by differential equation (4.1). The system matrices **A** and **B** are constant $n \times n$ and $n \times 1$ matrices, respectively. Operators **C** and **D** are the coefficients of $L(D)$ and $M(D)$, respectively.

It was previously seen from (3.39) that the system output is

$$y(t) = \langle \phi(t - t_0) | \mathbf{x}(t_0) \rangle + \int_{t_0}^{t} h(t - \xi) u(\xi) d\xi$$

The scalar product above identifies the zero-input response as

$$\langle \phi | \mathbf{x} \rangle = \varphi_1 x_1 + \varphi_2 x_2 + \dots + \varphi_n x_n$$

Using representation (4.33) and (4.34) equation (4.31) can be written as

$$Y(s) = H(s)U(s) + \sum_{i=1}^{n} \frac{s^{n-i}}{L(s)} x_i \tag{4.39}$$

The components x_i of the state vector are identified as the coefficients of the respective s^{n-i} in (4.33). Applying the inverse Laplace transform to both sides of (4.39) and comparing the results with (3.39) the zero-input responses are

$$\varphi_i = \mathcal{L}^{-1} \left\{ \frac{s^{n-i}}{L(s)} \right\} \qquad i = 1, 2, \dots, n \tag{4.40}$$

The responses $\varphi_1, \varphi_2, \dots, \varphi_n$ are linearly independent and constitute a set of basis functions for the system.

4.5 SYSTEM REALIZATION AND EQUIVALENCE

Given a linear system characterized by a single differential equation of the form (4.1) one can always construct (realize) an *equivalent* system comprised of an interconnection of adders, scalors, integrators and differentiators. In establishing the meaning of the equivalent system it is assumed that the reader is familiar with the analytic definitions of adders and scalors. Thus, further elaboration on these two terms is not warranted. For our purposes we will proceed to define integrators and differentiators. An integrator is a system whose input-output relation admits to the form

$$(D + a)y = u \tag{4.41}$$

where the constant a is not necessarily real. The integrator is denoted by $1/(D + a)$. The state of an integrator at time t is proportional to its output:

$$x(t) = ky(t)$$

where k is the constant of proportionality.

A differentiator is a system whose input-output relation admits to the form

$$y = (D + b)u \tag{4.42}$$

where b is also a constant not necessarily real. The differentiator is denoted by $(D + b)$. The state of a differentiator at time t is proportional to its input:

$$x(t) = k'u(t)$$

A vector $\mathbf{x}$ whose components are the outputs of the integrators and the inputs to the differentiators qualifies as a state vector of the equivalent system. (It is immediately obvious that by letting the output of an integrator also serve as the input of a differentiator the dimensionality of $\mathbf{x}$ for the equivalent system is reduced. Although a dimensional reduction of $\mathbf{x}$ is highly desirable from a computational point of view an arbitrary reduction or assignment of state components may prove to be futile in providing for a solution of the input-output-state or canonical equations.) *As a general rule the procedure for associating a state vector with a system comprised of adders, scalors, integrators and differentiators involves first assigning a component of* $\mathbf{x}$ *to the output of each integrator followed by assigning a component of* $\mathbf{x}$ *to the input of each differentiator not connected to the output of an integrator through a scalor.* This rule is exemplified diagramatically in Fig. 4.1, which shows integrators I_1 and I_2 interconnected with differentiators D_1 and D_2. The adders are denoted by the + signs and the scalors by the constants k_1 and k_2. State vector components x_1 and x_2 are assigned to the outputs of integrators I_1 and I_2, respectively. The remaining component, x_3, is assigned to the input of differentiator D_1. Note that the input to differentiator D_2 is directly connected to the output of integrator I_1. Thus no component is assigned D_2. The state vector $\mathbf{x}$ consists of components x_1, x_2 and x_3.

Another observation, not quite as apparent as the one cited above, involves virtual removal of the integrators and differentiators, thus reducing the system

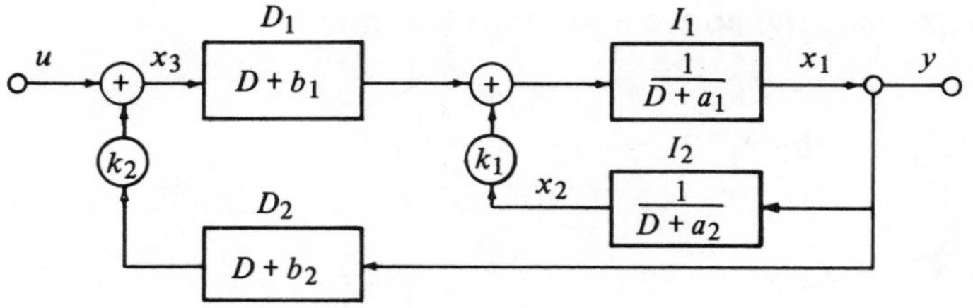

Figure 4.1 Associating state vector with system of adders, scalors, integrators and differentiators

to a memoryless interconnection of adders and scalors. Letting the output of the ith integrator be x_i we see from (4.41) that the input to the ith integrator is equal to $(D + a_i)x_i$. Similarly, letting x_j be the input to the jth differentiator by (4.42) the output of the jth differentiator is equal to $(D + b_j)x_j$. It can be reasoned that for an interconnection of adders, scalors, integrators and differentiators the input and output of the system will remain unaltered if the ith integrator is removed and replaced by an input x_i applied to the terminal to which the output of the ith integrator is connected. The term $(D + a_i)x_i$ plays the role of a suppressed output at the input terminal of the ith integrator. In like fashion the system output and input remains unaltered on removing the jth differentiator and replacing it by an input $(D + b_j)x_j$ applied to the terminal to which the output of the jth differentiator is connected. x_j plays the role of a suppressed output at the input terminal of the jth differentiator. On applying the above reasoning to the interconnection of Fig. 4.1 we see the system can be reduced to a series of segregated memoryless interconnections of adders and scalors, as exemplified in Fig. 4.2. From Fig. 4.2 we obtain by inspection the simultaneous equations

$$k_2(D + b_2)x_1 = x_3 - u$$

$$(D + b_1)x_3 - (D - a_1)x_1 = -k_1 x_2$$

$$(D + a_2)x_2 = x_1$$

$$y = x_1$$

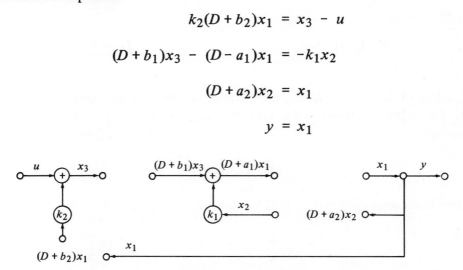

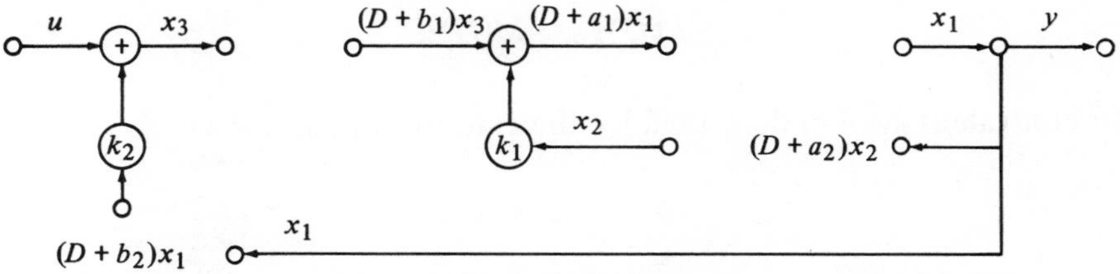

Figure 4.2 Memoryless system.

Upon rearrangement the above equations become

$$\dot{x}_1 = -b_2 x_1 + \frac{1}{k_2} x_3 - \frac{1}{k_2} u$$

$$\dot{x}_2 = x_1 - a_2 x_2$$

$$\dot{x}_3 = (a_1 - b_2) x_1 - k_1 x_2 + \left(\frac{1}{k_2} - b_1\right) x_3 - \frac{1}{k_2} u$$

$$y = x_1 \tag{4.43}$$

which can be written in canonical form as

$$\dot{\mathbf{x}} = \mathbf{A}\mathbf{x} + \mathbf{B}u$$

$$\mathbf{y} = \mathbf{C}\mathbf{x} + \mathbf{D}u$$

where

$$\mathbf{A} = \begin{bmatrix} -b_2 & 0 & \frac{1}{k_2} \\ 1 & -a_2 & 0 \\ (a_1 - b_2) & -k_1 & \left(\frac{1}{k_2} - b_1\right) \end{bmatrix} \qquad \mathbf{B} = \begin{bmatrix} -\frac{1}{k_2} \\ 0 \\ -\frac{1}{k_2} \end{bmatrix}$$

$$\mathbf{C} = [1 \quad 0 \quad 0] \qquad \mathbf{D} = 0$$

Equations (4.43) are the desired state equations for the equivalent system of Fig. 4.1. Thus, the relationship between the system described by the differential equation

$$L(D)y = M(D)u$$

and an equivalent system described by the canonical equations

$$\dot{\mathbf{x}} = \mathbf{A}\mathbf{x} + \mathbf{B}u$$

$$\mathbf{y} = \mathbf{C}\mathbf{x} + \mathbf{D}u$$

comes into sharper focus. Given any proper time-invariant system of finite order characterized by the above canonical equations, one can always structure an equivalent system characterized by a single (or set of) differential equation(s) of the form (4.1).

The above technique is but one of many means to *realize* a system described by an equation of form (4.1). In summary the technique is as follows: Given a system characterized by the relation

$$L(D)y = M(D)u$$

an *equivalent* system can be constructed of adders, scalors, differentiators and integrators. The equivalent system is a realization of the given system. A state vector $\mathbf{x}$ can be associated with the equivalent system. Since the two systems are equivalent it follows that $\mathbf{x}$ is also a state vector for the given system. Similarly, the state equations for the equivalent system may be regarded as the state equations of the given system.

4.6 METHOD OF PARTIAL FRACTIONS

As a special case of the realization technique we examine the methods of expansion by partial-fractions. Our interest in this method is in the fact that it yields state equations in which the $\mathbf{A}$ matrix is in diagonal form, thus the eigenvalues (or characteristic roots) associated with the system are readily identified. This, of course, requires factorization of $L(D)$, which in certain cases may prove to be a disadvantage of the technique. It is desired that $L(D)$ be factored and put into the form

$$L(D) = (D - \lambda_1)^{\alpha_1}(D - \lambda_2)^{\alpha_2} \dots (D - \lambda_q)^{\alpha_q} \tag{4.44}$$

where

$$L(D) = \ell_n D^n + \ell_{n-1} D^{n-1} + \dots + \ell_0$$

$$\alpha_1 + \alpha_2 + \dots + \alpha_q = n$$

The λ's are the distinct roots or zeros of $L(D)$ and the α's are their respective multiplicities.

To illustrate the partial fraction techniques we will analyze two examples: the case of simple zeros and the case of multiple zeros. For the case of simple zeros consider the system of integrators shown in Fig. 4.3. We have the conditions $\alpha_1 = \alpha_2 = \dots = \alpha_q = 1$, and $c_1, c_2, \dots$ are constants.

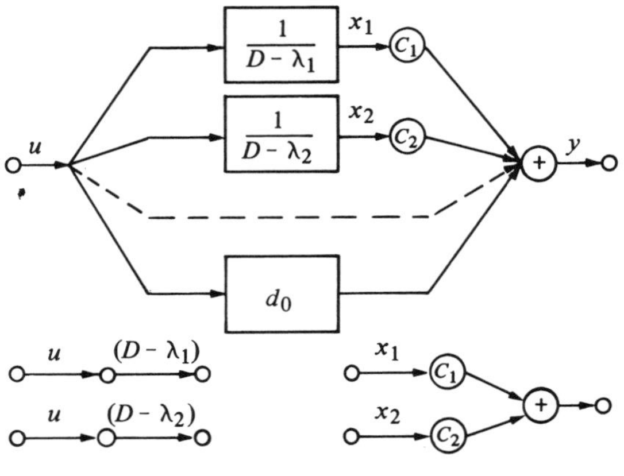

Figure 4.3 System of simple zeros.

Using (4.44) the form of the input-output relation is

$$(D-\lambda_1)(D-\lambda_2)(D-\lambda_3)\ldots(D-\lambda_n)y = M(D)u \tag{4.45}$$

Let the system of simple zeros be proper, i.e., the degree of $L(D)$ is greater than the degree of $M(D)$, and $(D-\lambda_i)$, where $i = 1,2,\ldots n$, is not a factor of $M(D)$. Clearly, by (4.45) the transfer function for the system having simple zeros is of the form

$$H(s) = \frac{M(s)}{(s-\lambda_1)(s-\lambda_2)\ldots(s-\lambda_n)} \tag{4.46}$$

On expanding $H(s)$ in terms of partial fractions we have

$$H(s) = \frac{c_1}{s-\lambda_1} + \frac{c_2}{s-\lambda_2} + \ldots + \frac{c_n}{s-\lambda_n} + d_0 \tag{4.47}$$

where the constants c_i are dependent on the λ_i and the coefficients of $M(D)$. The term d_0 is equal to zero if the degree of $M(D)$ does not exceed $n-1$.

Digressing temporarily, we verify that (4.47) is the transfer function for the system in Fig. 4.3. The Laplace transform of the state equations

$$\dot{\mathbf{x}} = \mathbf{Ax} + \mathbf{Bu}$$

$$\mathbf{y} = \mathbf{Cx} + \mathbf{Du}$$

where $\mathbf{A}$, $\mathbf{B}$, $\mathbf{C}$ and $\mathbf{D}$ are constant matrices, gives

$$s\mathbf{IX}(s) - \mathbf{x}(0) = \mathbf{AX}(s) + \mathbf{BU}(s) \tag{4.48}$$

$$\mathbf{Y}(s) = \mathbf{C}\mathbf{X}(s) + \mathbf{D}\mathbf{U}(s) \tag{4.49}$$

Solving the dynamical equation (4.48) for $\mathbf{X}(s)$ results in

$$\mathbf{X}(s) = (s\mathbf{I} - \mathbf{A})^{-1}\mathbf{x}(0) + (s\mathbf{I} - \mathbf{A})^{-1}\mathbf{B}\mathbf{U}(s)$$

which, when substituted in (4.49), yields for the output

$$\mathbf{Y}(s) = \mathbf{C}(s\mathbf{I} - \mathbf{A})^{-1}\mathbf{x}(0) + [\mathbf{C}(s\mathbf{I} - \mathbf{A})^{-1}\mathbf{B} + \mathbf{D}]\mathbf{U}(s) \tag{4.50}$$

The output in (4.50) identifies the system zero-state response and zero-input response in the frequency domain. We designate each respectively as $\mathbf{Z}(0;\mathbf{U}(s))$ and $\mathbf{Z}(\mathbf{X}(s);0)$. Letting the state $\mathbf{x}(0)$ be the xero state the system response is, from (4.50),

$$\mathbf{Z}(0;\mathbf{U}(s)) = [\mathbf{C}(s\mathbf{I} - \mathbf{A})^{-1}\mathbf{B} + \mathbf{D}]\mathbf{U}(s) \tag{4.51}$$

For zero input the system response is

$$\mathbf{Z}(\mathbf{X}(s);0) = \mathbf{C}(s\mathbf{I} - \mathbf{A})^{-1}\mathbf{x}(0) \tag{4.52}$$

As was specified earlier the matrix $\mathbf{D}$ is zero for a proper differential system. Letting the input to the system be the impulse $\delta(t)$, and since $\mathcal{L}\{\delta(t)\} = 1$, the transfer function is identified from (4.51) as

$$\mathbf{H}(s) = \mathbf{C}(s\mathbf{I} - \mathbf{A})^{-1}\mathbf{B} \tag{4.53}$$

In comparing (4.53) with (4.52) it is seen that for the simple case chosen—a proper system of simple zeros—the transfer function is related to the zero input response as

$$\mathbf{H}(s) = \mathbf{Z}(\mathbf{X}(s);\mathbf{B})$$

i.e., the transfer function $\mathbf{H}(s)$ is the Laplace transform of the zero input response starting in state $\mathbf{x}(0) = \mathbf{B}$. Assuming the matrix $(s\mathbf{I} - \mathbf{A})$ is nonsingular the matrix identity for the inverse operator can be written as

$$(s\mathbf{I} - \mathbf{A})^{-1} = \frac{\operatorname{adj}(s\mathbf{I} - \mathbf{A})}{|s\mathbf{I} - \mathbf{A}|}$$

where $|s\mathbf{I} - \mathbf{A}|$ and $\operatorname{adj}(s\mathbf{I} - \mathbf{A})$ are the determinant and adjoint of $(s\mathbf{I} - \mathbf{A})$, respectively. From (4.53) the transfer function becomes

$$\mathbf{H}(s) = \frac{\mathbf{C}[\mathrm{adj}(s\mathbf{I} - \mathbf{A})]\mathbf{B}}{|s\mathbf{I} - \mathbf{A}|} \tag{4.54}$$

By inspection the state equations of the system of simple zeros shown in Fig. 4.3 are

$$\dot{x}_1 = \lambda_1 x_1 + u$$

$$\dot{x}_2 = \lambda_2 x_2 + u$$

$$\cdots\cdots\cdots\cdots$$

$$\dot{x}_n = \lambda_n x_n + u$$

$$y = c_1 x_1 + c_2 x_2 + \ldots + c_n x_n \tag{4.55}$$

from which the system matrices are

$$\mathbf{A} = \begin{bmatrix} \lambda_1 & 0 & \ldots \\ 0 & \lambda_2 & \ldots \\ \cdots & \cdots & \cdots \\ 0 & \ldots & \lambda_n \end{bmatrix} \qquad \mathbf{B} = \begin{bmatrix} 1 \\ 1 \\ \cdots \\ 1 \end{bmatrix} \qquad \mathbf{C} = [c_1\, c_2 \ldots c_n] \tag{4.56}$$

Substituting the expressions for matrices **A**, **B** and **C** in (4.54) yields

$$H(s) = \frac{c_1}{s - \lambda_1} + \frac{c_2}{s - \lambda_2} + \ldots + \frac{c_n}{s - \lambda_n}$$

thus, verifying (4.47).

In accordance with the above analysis we assert that for a proper system of simple zeros characterized by an input-output relation of the form (4.45)

$$(D - \lambda_1)(D - \lambda_2) \ldots (D - \lambda_n) = M(D)u$$

where the various λ_i are distinct and no term $(D - \lambda_i)$ is a factor of $M(D)$, the transfer function $H(s)$ can be represented in partial fraction form (4.47):

$$H(s) = \frac{c_1}{s - \lambda_1} + \frac{c_2}{s - \lambda_2} + \ldots + \frac{c_n}{s - \lambda_n}$$

The system as shown in Fig. 4.3 is a realization of the system represented by equation (4.45). The vector

$$\mathbf{x}(t) = (x_1, x_2, \ldots x_n)$$

defined by (4.55) as

$$\dot{x} = \lambda_i x_i + u \qquad i = 1,2,\ldots,n$$

qualifies as the system state vector. Equations (4.56) are the corresponding state equations for the system.

In extending the above discussion to the case of multiple zeros we first consider, for simplicity, the case where $L(D)$ has one multiple zero λ_1 of order α. The expression for $L(D)$ becomes

$$L(D)y = [(D - \lambda_1)^\alpha (D - \lambda_{\alpha+1}) \ldots (D - \lambda_n)]y$$
$$= M(D)u \tag{4.57}$$

The corresponding expression for $H(s)$ can be written as

$$H(s) = \frac{M(s)}{(s - \lambda_1)^\alpha (s - \lambda_{\alpha+1}) \ldots (s - \lambda_n)} \tag{4.58}$$

Expanding $H(s)$ into partial fractions gives

$$H(s) = \frac{c_1}{(s - \lambda_1)^\alpha} + \frac{c_2}{(s - \lambda_1)^{\alpha-1}} + \ldots + \frac{c_\alpha}{(s - \lambda_1)}$$
$$+ \frac{c_{\alpha+1}}{(s - \lambda_{\alpha+1})} + \ldots + \frac{c_n}{(s - \lambda_n)} + d_0 \tag{4.59}$$

By (4.59) it is seen that, in essence, *one* multiple zero of order α has the equivalent effect of adding α integrators to *one* of the integrator circuits. This is shown graphically[1] in Fig. 4.4. (Obviously, for the case of more than one multiple zero, say m multiple zeros each of order $\alpha, \beta, \gamma, \ldots, \xi$, respectively, the equivalent effect would be to add α integrators to the respective circuit labeled a, β integrators to the respective circuit $b, \ldots$ ξ integrators to the respective circuit labeled m.) Choosing the output of each integrator as an element of the state vector $\mathbf{x} = (x_1, x_2, \ldots, x_n)$ the state equations are, by inspection,

1. From Zedeh and Desoer, Linear System Theory, McGraw-Hill, New York 1963.

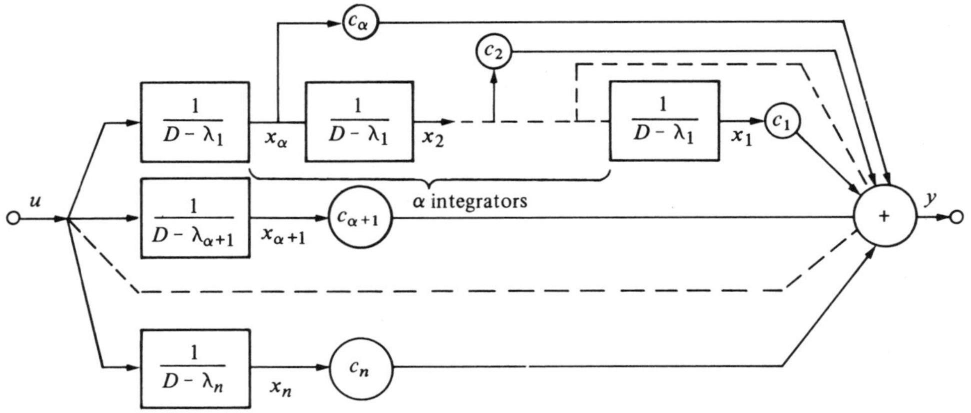

Figure 4.4 System of multiple zeros.

$$
\begin{aligned}
\dot{x}_1 &= \lambda_1 x_1 + x_2 \\
&\dots\dots\dots\dots \\
\dot{x}_{\alpha-1} &= \lambda_1 x_{\alpha-1} + x_\alpha \\
\dot{x}_\alpha &= \lambda_1 x_\alpha + \alpha \\
&\dots\dots\dots\dots \\
\dot{x}_n &= \lambda_n x_n + u \\
y &= c_1 x_1 + c_2 x_2 + \dots + c_n x_n + d_0
\end{aligned} \tag{4.60}
$$

wherein it is clear that this choice of **x** qualifies as a system state vector. Hence, for a proper system characterized by the input-output relation

$$(D - \lambda_1)^\alpha (D - \lambda_{\alpha+1}) \dots (D - \lambda) y = M(D) u$$

where the various λ_i are distinct and no term $(D - \lambda_i)$ is a factor of $M(D)$, the transfer function $H(s)$ can be represented in partial fraction form (4.59). The system of Fig. 4.4 is a realization of (4.57). The corresponding state equations in canonical form are, from (4.60),

$$\dot{\mathbf{x}} = \mathbf{A}\mathbf{x} + \mathbf{B}\mathbf{u}$$

$$\mathbf{y} = \langle \mathbf{C} | \mathbf{x} \rangle + d_0 u$$

where

$$\mathbf{A} = \begin{bmatrix} \lambda_1 & 1 & 0 & \dots & 0 \\ 0 & \lambda_1 & 1 & \dots & 0 \\ 0 & 0 & \lambda_1 & 1 \dots & 0 \\ \dots & \dots & \dots & \dots & \dots \\ 0 & 0 & 0 & \dots & \lambda_n \end{bmatrix} \qquad \mathbf{B} = \begin{bmatrix} 0 \\ \vdots \\ 0 \\ \vdots \\ 1 \end{bmatrix}$$

$$\mathbf{C} = [c_1 \; c_2 \; \dots \; c_n]$$

5
Solutions to the Canonical Equations

5.1 FIXED CONTINUOUS-TIME SYSTEMS: TIME DOMAIN ANALYSIS

The differential equations governing the behavior of linear continuous time systems were established in (2.26) and (2.27) as

$$\dot{\mathbf{x}}(t) = \mathbf{A}(t)\mathbf{x}(t) + \mathbf{B}(t)\mathbf{u}(t)$$

$$\mathbf{y}(t) = \mathbf{C}(t)\mathbf{x}(t) + \mathbf{D}(t)\mathbf{u}(t)$$

For fixed systems the matrices **A**, **B**, **C** and **D** are fixed, i.e., they are independent of time. In determining the time-dependent solutions to (2.26) and (2.27) for fixed systems we first consider the homogeneous form of (2.26) where $u(t) = 0$. Equation (2.26) becomes

$$\frac{d}{dt}\mathbf{x}(t) = \mathbf{A}\mathbf{x}(t) \tag{5.1}$$

where the matrix $\mathbf{A}$ is a constant matrix. Solving (5.1) for $\mathbf{x}(t)$ between the time limits t_0 and t we have

$$\int_{t_0}^{t} \frac{d\mathbf{x}(t)}{\mathbf{x}(t)} = \mathbf{A}\int_{t_0}^{t} dt$$

$$\ln \frac{\mathbf{x}(t)}{\mathbf{x}(t_0)} = \mathbf{A}(t - t_0)$$

$$\mathbf{x}(t) = e^{\mathbf{A}(t-t_0)}\mathbf{x}(t_0) \tag{5.2}$$

Upon comparing (5.2) with (3.47), wherein

$$\mathbf{x}(t) = \Phi(t - t_0)\mathbf{x}(t_0)$$

it is seen that the state transition matrix can be represented as

$$\begin{aligned}\Phi(t - t_0) &= e^{\mathbf{A}t} \\ &= \mathbf{I} + \mathbf{A}(t - t_0) + \mathbf{A}^2 \frac{(t - t_0)^2}{2!} + \ldots \end{aligned} \tag{5.3}$$

Thus, by (5.3), $\Phi(t)$ is also referred to as the *fundamental matrix*.

The complete time-dependent solution to (2.26), where $\mathbf{A}$ and $\mathbf{B}$ are constant matrices, is readily obtained through the method of *variation of the parameter*. Assume the solution to (2.26) is

$$\mathbf{x}(t) = e^{\mathbf{A}(t-t_0)}\mathbf{f}(t) \tag{5.4}$$

where $\mathbf{f}(t)$ is to be determined. The time derivative of solution (5.4) gives

$$\dot{\mathbf{x}}(t) = \mathbf{A}e^{\mathbf{A}(t-t_0)}\mathbf{f}(t) + e^{\mathbf{A}(t-t_0)}\dot{\mathbf{f}}(t) \tag{5.5}$$

Substituting (5.4) into (2.26) and using identity (5.5) gives

$$\begin{aligned}\dot{\mathbf{x}} - \mathbf{A}e^{\mathbf{A}(t-t_0)}\mathbf{f}(t) &= \mathbf{B}\mathbf{u}(t) \\ &= e^{\mathbf{A}(t-t_0)}\dot{\mathbf{f}}(t)\end{aligned} \tag{5.6}$$

Multiplying each side of (5.6) from the left by $e^{-\mathbf{A}(t-t_0)}$ and integrating between the limits $(-\infty, t)$, assuming $\mathbf{f}(-\infty) = 0$, we have

$$\int_{-\infty}^{t} e^{\mathbf{A}(\xi - t_0)}\mathbf{Bu}(\xi)d\xi = \int_{-\infty}^{t} d\mathbf{f}(\xi)$$

$$\mathbf{f}(t) = \int_{-\infty}^{t} e^{-\mathbf{A}(\xi - t_0)}\mathbf{Bu}(\xi)d\xi$$

which, when substituted in solution (5.4), gives

$$\mathbf{x}(t) = e^{\mathbf{A}(t-t_0)} \int_{-\infty}^{t} e^{-\mathbf{A}(\xi - t_0)}\mathbf{Bu}(\xi)d\xi$$

$$= e^{\mathbf{A}(t-t_0)} \int_{-\infty}^{t_0} e^{-\mathbf{A}(\xi - t_0)}\mathbf{Bu}(\xi)d\xi + \int_{t_0}^{t} e^{\mathbf{A}(t-\xi)}\mathbf{Bu}(\xi)d\xi \quad (5.7)$$

Evaluating (5.7) at $t = t_0$ gives the initial state as

$$\mathbf{x}(t_0) = \int_{-\infty}^{t_0} e^{-\mathbf{A}(\xi - t_0)}\mathbf{Bu}(\xi)d\xi$$

Therefore, solution (5.7) becomes

$$\mathbf{x}(t) = e^{\mathbf{A}(t-t_0)}\mathbf{x}(t_0) + \int_{t_0}^{t} e^{\mathbf{A}(t-\xi)}\mathbf{Bu}(\xi)d\xi \quad (5.8)$$

$$= \Phi(t - t_0)\mathbf{x}(t_0) + \int_{t_0}^{t} \Phi(t - \xi)\mathbf{Bu}(\xi)d\xi \quad (5.9)$$

For fixed systems it is convenient to establish $t_0 = 0$, in which case equations (5.8) and (5.9) become

$$\mathbf{x}(t) = e^{\mathbf{A}t}\mathbf{x}(0) + \int_{0}^{t} e^{\mathbf{A}(t-\xi)}\mathbf{Bu}(\xi)d\xi \quad (5.10)$$

$$= \Phi(t)\mathbf{x}(0) + \int_{0}^{t} \Phi(t - \xi)\mathbf{Bu}(\xi)d\xi \quad (5.11)$$

Finally, on substituting (5.11) into (2.27) the input-output-state relation for a continuous-time system is

$$\mathbf{y}(t) = \mathbf{C}\left[\Phi(t)\mathbf{x}(0) + \int_0^t \Phi(t-\xi)\mathbf{B}\mathbf{u}(\xi)d\xi\right] + \mathbf{D}\mathbf{u} \tag{5.12}$$

Equations (5.11) and (5.12) are the system state equations. It is clear from (5.11) that the system state at time t can be determined if the system state at some previous time t_0 ($t_0 < t$) is known and if the input $\mathbf{u}(t)$ is known. The manner in which the initial state is "transformed" is characterized by the make-up of Φ and how the input is applied, i.e., by the matric operators **A** and **B**. The system output (5.12) reflects a dependency on all four operators.

5.2 FIXED CONTINUOUS-TIME SYSTEMS: FREQUENCY DOMAIN ANALYSIS

For fixed continuous-time systems the differential equations to be solved are

$$\dot{\mathbf{x}} = \mathbf{A}\mathbf{x} + \mathbf{B}\mathbf{u}$$

$$\mathbf{y} = \mathbf{C}\mathbf{x} + \mathbf{D}\mathbf{u}$$

where **A**, **B**, **C** and **D** are constant matrices. From previous discussions the method of Laplace transforms was seen to be a convenient method for solving equations of this form. It was established in (4.48) and (4.49) that the Laplace transform of both sides of the above equations gives, respectively,

$$s\mathbf{X}(s) - \mathbf{x}(0) = \mathbf{A}\mathbf{X}(s) + \mathbf{B}\mathbf{U}(s)$$

$$\mathbf{Y}(s) = \mathbf{C}\mathbf{X}(s) + \mathbf{D}\mathbf{U}(s)$$

From (4.48)

$$\mathbf{X}(s) = (s\mathbf{I} - \mathbf{A})^{-1}\mathbf{x}(0) + (s\mathbf{I} - \mathbf{A})^{-1}\mathbf{B}\mathbf{U}(s) \tag{5.13}$$

$$= \begin{bmatrix} X_1(s) \\ X_2(s) \\ \vdots \\ X_n(s) \end{bmatrix} = \begin{bmatrix} \mathcal{L}\{x_1(t)\} \\ \mathcal{L}\{x_2(t)\} \\ \vdots \\ \mathcal{L}\{x_n(t)\} \end{bmatrix} \tag{5.14}$$

which, when substituted in (4.49), resulted in the input-output equation

$$\mathbf{Y}(s) = \mathbf{C}(s\mathbf{I} - \mathbf{A})^{-1}\mathbf{x}(0) + [\mathbf{C}(s\mathbf{I} - \mathbf{A})^{-1}\mathbf{B} + \mathbf{D}]\mathbf{U}(s)$$

$$= \begin{bmatrix} Y_1(s) \\ Y_2(s) \\ \vdots \\ Y_n(s) \end{bmatrix} = \begin{bmatrix} \mathcal{L}\{y_1(t)\} \\ \mathcal{L}\{y_2(t)\} \\ \vdots \\ \mathcal{L}\{y_n(t)\} \end{bmatrix} \tag{5.15}$$

The zero-state response, i.e., $\mathbf{x}(0) = 0$, identifies the *transfer function matrix* $\mathbf{H}(s)$ in (4.50) as

$$\mathbf{H}(s) = \mathbf{C}(s\mathbf{I} - \mathbf{A})^{-1}\mathbf{B} + \mathbf{D} \tag{5.16}$$

$$= \begin{bmatrix} H_{11}(s) & H_{12}(s) & \dots & H_{1n}(s) \\ \dots & \dots & \dots & \dots \\ H_{n1}(s) & H_{n2}(s) & \dots & H_{nn}(s) \end{bmatrix} \tag{5.17}$$

Writing the zero-state response as

$$\mathbf{Y}(s) = \mathbf{H}(s)\mathbf{U}(s) \tag{5.18}$$

the ith component of the transform vector $\mathbf{Y}(s)$ is

$$Y_i(s) = H_{i1}U_1(s) + H_{i2}U_2(s) + \dots + H_{in}U_n(s) \tag{5.19}$$

It is apparent from (5.19) that $H_{ij}(s)$ is the transfer function between input $u_j(t)$ and output $y_i(t)$. In Section 3.4 $h_{ij}(t)$ was identified as the response at the ith output terminal due to a unit impulse applied at the jth input terminal. We conclude, therefore, that

$$H_{ij}(s) = \mathcal{L}\{h_{ij}(t)\} \tag{5.20}$$

Equation (5.10) gave the general time-domain solution for a fixed system as

$$\mathbf{x}(t) = \Phi(t)\mathbf{x}(0) + \int_0^t \Phi(t - \xi)\mathbf{B}\mathbf{u}(\xi)d\xi$$

where, for convenience, $t_0 = 0$. Applying the theorem for the Laplace transform of a convolution the Laplace transform of the above equation is

$$\mathbf{X}(s) = \hat{\Phi}(s)\mathbf{x}(0) + \hat{\Phi}(s)\mathbf{B}\mathbf{X}(s) \tag{5.21}$$

where

$$\hat{\Phi}(s) = \mathcal{L}\{\Phi(t)\} \tag{5.22}$$

$\Phi(t)$ has previously been identified as the fundamental matrix of the system. Comparing (5.21) with (5.13) the Laplace transform of the fundamental matrix is

$$\hat{\Phi}(s) = (s\mathbf{I} - \mathbf{A})^{-1} \tag{5.23}$$

or

$$e^{\mathbf{A}t} = \Phi(t) = \mathcal{L}^{-1}\{(s\mathbf{I} - \mathbf{A})^{-1}\} \tag{5.24}$$

Thus, the fundamental matrix $\Phi(t)$ is the inverse Laplace transform of the matrix $(s\mathbf{I} - \mathbf{A})^{-1}$. Accordingly, $\hat{\Phi}(s)$ is referred to as the *characteristic frequency matrix* of the system.

Using the matrix identity for an inverse operator equation (5.23) can be written as

$$\hat{\Phi}(s) = (s\mathbf{I} - \mathbf{A})^{-1} = \frac{\operatorname{adj}(s\mathbf{I} - \mathbf{A})}{|s\mathbf{I} - \mathbf{A}|} \tag{5.25}$$

where $|s\mathbf{I} - \mathbf{A}|$ and $\operatorname{adj}(s\mathbf{I} - \mathbf{A})$ are the determinant and adjoint of $(s\mathbf{I} - \mathbf{A})$, respectively. For a system of order n, $\mathbf{A}$ is an $n \times n$ matrix. Therefore the determinant $|s\mathbf{I} - \mathbf{A}|$ is a polynomial in s of degree n, which can be written as

$$\begin{aligned} |s\mathbf{I} - \mathbf{A}| &= s + a_1 s^{n-1} + a_2 s^{n-2} + \ldots + a_{n-1}s + a_n \\ &= (s - \lambda_1)^{\alpha_1}(s - \lambda_2)^{\alpha_2}\ldots(s - \lambda_n)^{\alpha_n} \\ &= p(s) \end{aligned} \tag{5.26}$$

Polynomial (5.26) is the *characteristic polynomial* of the system. Its zeros $\lambda_1, \lambda_2, \ldots, \lambda_n$ (those values of s for which $|s\mathbf{I} - \mathbf{A}| = 0$) are the *characteristic roots* or *eigenvalues* of the system. Accordingly, the equation

$$|s\mathbf{I} - \mathbf{A}| = 0 \tag{5.27}$$

is called the *characteristic equation* of the system.

For the simple case where n is 2 or 3 expansion of adj$(s\mathbf{I} - \mathbf{A})$ in (5.25) by cofactors suffices to determine $\hat{\Phi}(s)$. Let $(s\mathbf{I} - \mathbf{A})_{ij}$ denote the $(n-1) \times (n-1)$ submatrix of $(s\mathbf{I} - \mathbf{A})$ formed by deleting row i and column j. The scalar

$$c_{ij} = (-1)^{i+j}|(s\mathbf{I} - \mathbf{A})_{ij}| \tag{5.28}$$

is the cofactor of the (i,j) element of $(s\mathbf{I} - \mathbf{A})$. The $n \times n$ matrix $(s\mathbf{I} - \mathbf{A})'$ formed by all the c_{ij} elements is the cofactor matrix of $(s\mathbf{I} - \mathbf{A})$:

$$\text{cofactor } (s\mathbf{I} - \mathbf{A}) = (s\mathbf{I} - \mathbf{A})' = \begin{bmatrix} c_{11} & c_{12} & \cdots & c_{1n} \\ c_{21} & c_{22} & \cdots & c_{2n} \\ \cdots & \cdots & \cdots & \cdots \\ c_{n1} & c_{n2} & \cdots & c_{nn} \end{bmatrix} \tag{5.29}$$

The adjoint of $(s\mathbf{I} - \mathbf{A})$ is the transpose of the cofactor matrix of $(s\mathbf{I} - \mathbf{A})$:

$$\text{adj}(s\mathbf{I} - \mathbf{A}) = \widetilde{(s\mathbf{I} - \mathbf{A})}' = \begin{bmatrix} c_{11} & c_{21} & \cdots & c_{n1} \\ c_{12} & c_{22} & \cdots & c_{n2} \\ \cdots & \cdots & \cdots & \cdots \\ c_{1n} & c_{2n} & \cdots & c_{nn} \end{bmatrix} \tag{5.30}$$

The tilde denotes the transpose.

In principle the above procedure provides for determining $\hat{\Phi}(s)$. However, for values of n greater than 3 this procedure becomes impractical since, essentially, $n^2(n-1) \times (n-1)!$ multiplications are involved. A more practical technique, which also lends itself to machine computation, involves extending the method of expanding a rational function to matrices whose elements are rational functions. We let

$$(s\mathbf{I} - \mathbf{A})^{-1} = \frac{\mathbf{B}(s)}{|s\mathbf{I} - \mathbf{A}|} = \frac{\mathbf{B}(s)}{p(s)} \tag{5.31}$$

where

$$\mathbf{B}(s) = \mathbf{B}_0 s^{n-1} + \mathbf{B}_1 s^{n-2} + \ldots + \mathbf{B}_{n-2} s + \mathbf{B}_{n-1} \tag{5.32}$$

and

$$p(s) = s + a_1 s^{n-1} + a_2 s^{n-2} + \ldots + a_n$$

The coefficients $\mathbf{B}_0, \mathbf{B}_1, \ldots, \mathbf{B}_{n-1}$ are constant $n \times n$ matrices, and $p(s)$ was previously defined in (5.26). Given an $n \times n$ matrix $\mathbf{A}$ the coefficients a_i of the polynomial $p(s)$ and the coefficients $\mathbf{B}_i$ of $\mathbf{B}(s)$ are determined by the following algorithms:[1]

$$a_i = -\frac{1}{i} \operatorname{tr}(\mathbf{B}_{i-1}\mathbf{A}) \qquad i = 1,2,\ldots,n \tag{5.33}$$

$$\mathbf{B}_i = \mathbf{B}_{i-1}\mathbf{A} + a_i\mathbf{I} \qquad i = 1,2,\ldots,n \tag{5.34}$$

The above algorithm leads to the following relations between the coefficients a_i and $\mathbf{B}_i$:

$$\begin{aligned}
a_1 &= -\operatorname{tr}(\mathbf{A}) & \mathbf{B}_0 &= \mathbf{I} \\
a_2 &= -\frac{1}{2}\operatorname{tr}(\mathbf{B}_1\mathbf{A}) & \mathbf{B}_1 &= \mathbf{B}_0\mathbf{A} + a_1\mathbf{I} \\
a_3 &= -\frac{1}{3}\operatorname{tr}(\mathbf{B}_2\mathbf{A}) & \mathbf{B}_2 &= \mathbf{B}_1\mathbf{A} + a_2\mathbf{I} \\
&\cdots\cdots & &\cdots\cdots \\
a_{n-1} &= -\frac{1}{n-1}\operatorname{tr}(\mathbf{B}_{n-2}\mathbf{A}) & \mathbf{B}_{n-1} &= \mathbf{B}_{n-2}\mathbf{A} + a_{n-1}\mathbf{I} \\
a_n &= -\frac{1}{n}\operatorname{tr}(\mathbf{B}_{n-1}\mathbf{A}) & 0 &= \mathbf{B}_{n-1}\mathbf{A} + a_n\mathbf{I}
\end{aligned} \tag{5.35}$$

Equations (5.35) are a more efficient procedure in computing $(s\mathbf{I} - \mathbf{A})^{-1}$. Only n matrix multiplications are required as opposed to $n^2(n-1)(n-1)!$ using the method of expansion by cofactors.

To complete the computation of $\Phi(t)$ in (5.24) we compute the inverse Laplace transform of $(s\mathbf{I} - \mathbf{A})^{-1}$. Equations (5.35) result in a matrix for $(s\mathbf{I} - \mathbf{A})^{-1}$ whose elements are rational functions. The inverse Laplace transform of a proper rational function (considered to be the ratio of two polynomials where the degree of the numerator is greater than that of the denominator) is most readily obtained by expanding the rational function into partial fractions. As an example let $F(s)$ be a proper rational function in s defined as

$$F(s) = N(s)/Q(s) \tag{5.36}$$

1. V. N. Faddeeva, "Computational Methods of Linear Algebra," Dover Publications, Inc., New York, 1959.

If the degree of $N(s)$ is greater than that of $Q(s)$, and if $F(s)$ has poles at $s = \lambda_1, \lambda_2, \ldots, \lambda_n$ and the order of the pole at $s = \lambda_j$ is α_j then (5.36) can be factored as

$$F(s) = \sum_{j=1}^{n} \sum_{i=1}^{\alpha_j} Y_{ij} \frac{1}{(s - \lambda_j)^i} \tag{5.37}$$

A hint of the structure of (5.37) was first seen in (4.58) and (4.59). Each nonrepeated factor $s - \lambda_\ell$ in the denominator of (5.37) results in a term

$$Y_{1\ell} \frac{1}{(s - \lambda_\ell)}$$

whereas each of the repeated factors $(s - \lambda_j)^{\alpha_j}$ give rise to α_j terms

$$Y_{\alpha_j j} \frac{1}{(s - \lambda_j)^{\alpha_j}} + Y_{\alpha_j - 1 j} \frac{1}{(s - \lambda_j)^{\alpha_j - 1}} + \ldots + Y_{1j} \frac{1}{(s - \lambda_j)}$$

For a simple first-order pole at $s = \lambda_\ell$ the coefficients Y_{ij} may be evaluated by the formula

$$Y_{1\ell} = \left. \frac{N(s)}{Q(s)/(s - \lambda_\ell)} \right|_{s=\lambda_\ell} \tag{5.38}$$

$$= \frac{N(\lambda_\ell)}{D(\lambda_\ell)} \tag{5.39}$$

where $D(\lambda_\ell)$ is the derivative of $Q(s)$ evaluated at $s = \lambda_\ell$. For a pole of order α_j at $s = \lambda_j$ we have

$$Y_{ij} = \left. \frac{1}{(\alpha_j - i)!} \frac{d^{\alpha_j - i}}{ds^{\alpha_j - i}} \left[\frac{N(s)}{Q(s)/(s - \lambda_j)^{\alpha_j}} \right] \right|_{s=\lambda_j} \tag{5.40}$$

where $i = 1, 2, \ldots, \alpha_j$.

To obtain the inverse transform of $F(s)$ consider the following: The Laplace transform of a time function of the form $f(t) = t^{n-1}/(n-1)!$ is

$$\mathcal{L}\left\{ \frac{t^{n-1}}{(n-1)!} \right\} = \frac{1}{s} \tag{5.41}$$

Secondly, the transform of the product of two functions, wherein one of the functions is an exponential, is

$$\mathcal{L}\{f(t)e^{-\lambda t}\} = F(s+\lambda) \tag{5.42}$$

Combining the ideas of (5.41) and (5.42) we can write

$$\mathcal{L}\left\{\frac{t^{n-1}}{(n-1)!}e^{-\lambda t}\right\} = \frac{1}{(s+\lambda)^n} \tag{5.43}$$

Applying the rationale of (5.43) to (5.37) the inverse Laplace transform of (5.37) can be written as

$$f(t) = \mathcal{L}^{-1}\{F(s)\}$$

$$f(t) = \mathcal{L}^{-1}\left\{\sum_{j=1}^{n}\sum_{i=1}^{\alpha_j} Y_{ij}\frac{1}{(s-\lambda_j)^i}\right\}$$

$$= \sum_{j=1}^{n} e^{\lambda_j t}\sum_{i=1}^{\alpha_j} Y_{ij}\frac{t^{i-1}}{(i-1)!} \tag{5.44}$$

Expression (5.44) prescribes a method for determining the inverse Laplace transform of each element of $\hat{\Phi}(s) = (s\mathbf{I} - \mathbf{A})^{-1}$, thereby determining $\Phi(t)$. The matrix form of $\Phi(t)$ can be established by expanding (5.37). We obtain

$$\hat{\Phi}(s) = \sum_{j=1}^{n'\leqslant n}\sum_{i=1}^{\alpha_j} \mathbf{Y}_{ij}\frac{1}{(s-\lambda_j)^i} \tag{5.45}$$

where $j = 1, 2, \ldots, n' \leqslant n$. The inverse Laplace transform of (5.45) gives

$$\Phi(t) = e^{\mathbf{A}t} = \sum_{j=1}^{n'\leqslant n} e^{\lambda_j t}\sum_{i=1}^{\alpha_j} \mathbf{Y}_{ij}\frac{t^{i-1}}{(i-1)!} \tag{5.46}$$

where (5.44) served as a guide to establish the general form of $\Phi(t)$.

To illustrate the use of equations (5.37) through (5.44) consider the inverse Laplace transform of

$$F(s) = \frac{-2s^4 - 11s - 16}{s^2(s+1)^2(s+2)} \tag{5.47}$$

For the term s^2 the multiplicity α_j is 2. The respective coefficients Y_{ij} determined by (5.40) are $Y_{11} = 2$ and $Y_{21} = -3$. Similarly, the coefficients Y_{ij} associated with the degenerate eigenvalue of $(s + 1)^2$ are $Y_{12} = 0$ and $Y_{22} = 3$. From (5.38) or (5.39) the term $(s + 2)$ has the coefficient $Y_{13} = -4$. Thus (5.47) can be written in form (5.37) as

$$F(s) = \frac{2}{s} - \frac{3}{s^2} + \frac{3}{(s+1)^2} - \frac{4}{(s+2)} \tag{5.48}$$

By (5.44) the inverse Laplace transform of (5.48) is

$$f(t) = 2 - 3t + 3te^{-t} - 4e^{-2t} \tag{5.49}$$

Clearly, the inversion of the $n \times n$ matrix $\hat{\Phi}(s)$, using the above method, requires n^2 elemental inversions.

5.3 FIXED DISCRETE-TIME SYSTEMS: TIME DOMAIN ANALYSIS

A linear *discrete-time* system of fixed form is governed by a system of *difference* equations of normal form[1], from (2.26) and (2.27),

$$\mathbf{x}(k+1) = \mathbf{A}\mathbf{x}(k) + \mathbf{B}\mathbf{u}(k) \tag{5.50}$$

$$\mathbf{y}(k) = \mathbf{C}\mathbf{x}(k) + \mathbf{D}\mathbf{u}(k) \tag{5.51}$$

where the index $k = 1,2,\ldots,\infty$ prescribes the discrete-time sequence and the matrices **A**, **B**, **C** and **D** are constant. The solution to (5.50) may be found directly by recursion. Substituting in (5.50) discrete values of k we have

$$\begin{aligned}
\mathbf{x}(1) &= \mathbf{A}\mathbf{x}(0) + \mathbf{B}\mathbf{u}(0) \\
\mathbf{x}(2) &= \mathbf{A}\mathbf{x}(1) + \mathbf{B}\mathbf{u}(1) \\
&= \mathbf{A}^2\mathbf{x}(0) + \mathbf{A}\mathbf{B}\mathbf{u}(0) + \mathbf{B}\mathbf{u}(1) \\
\mathbf{x}(3) &= \mathbf{A}\mathbf{x}(2) + \mathbf{B}\mathbf{u}(2) \\
&= \mathbf{A}^3\mathbf{x}(0) + \mathbf{A}^2\mathbf{B}\mathbf{u}(0) + \mathbf{A}\mathbf{B}\mathbf{u}(1) + \mathbf{B}\mathbf{u}(2) \\
&\cdots\cdots\cdots\cdots\cdots\cdots\cdots\cdots\cdots
\end{aligned}$$

$$\mathbf{x}(k) = \mathbf{A}^k\mathbf{x}(0) + \sum_{\ell=0}^{k-1} \mathbf{A}^{\ell}\mathbf{B}\mathbf{u}(k - \ell - 1) \tag{5.52}$$

1. See, for example, Freeman, H., Discrete-Time Systems, John Wiley and Sons, New York (1965).

or

$$\mathbf{x}(k) = \mathbf{A}^k\mathbf{x}(0) + \sum_{\ell=0}^{k-1} \mathbf{A}^{k-\ell-1}\mathbf{B}\mathbf{u}(\ell) \tag{5.53}$$

The *fundamental matrix* for a discrete-time system becomes, from (5.52) or (5.53),

$$\Phi(k) = \mathbf{A}^k \tag{5.54}$$

where

$$\mathbf{A}^k = \mathbf{A}_1\mathbf{A}_2 \ldots \mathbf{A}_k$$

$$\mathbf{A}_1 = \mathbf{A}_2 = \ldots \mathbf{A}_k = \mathbf{A} \tag{5.55}$$

In terms of the fundamental matrix $\Phi(k)$ the state equation (5.52) or (5.53) becomes, respectively,

$$\mathbf{x}(k) = \Phi(k)\mathbf{x}(0) + \sum_{\ell=0}^{k-1} \Phi(\ell)\mathbf{B}\mathbf{u}(k-\ell-1) \tag{5.56}$$

or

$$\mathbf{x}(k) = \Phi(k)\mathbf{x}(0) + \sum_{\ell=0}^{k-1} \Phi(k-\ell-1)\mathbf{B}\mathbf{u}(k) \tag{5.57}$$

Substituting (5.56) into (5.51) the output equation becomes

$$\mathbf{y}(k) = \mathbf{C}\Phi(k)\mathbf{x}(0) + \sum_{\ell=0}^{k-1} \mathbf{C}\Phi(\ell)\mathbf{B}\mathbf{u}(k-\ell-1) + \mathbf{D}\mathbf{u}(k) \tag{5.58}$$

An alternate expression for the output is readily derived using the state vector form (5.57). Solutions (5.56) and (5.58) are the discrete-time analogs of (5.11) and (5.12), respectively. The corresponding formulation of the discrete-time state equations is given in Appendix C.

5.4 FIXED CONTINUOUS-TIME SYSTEMS: DISCRETE INPUTS

In many continuous-time systems the input is often a sampled signal, thus characterizing the input as discrete. There are two kinds of sampled signals (and associated discreteness) we will consider: (1) the piece-wise constant signals where the sampling function is a pulse of constant amplitude over a time interval (t_k, t_{k+1}), i.e.,

$$\mathbf{u}(t) = \mathbf{u}(t_k) \qquad t_k < t \leqslant t_{k+1} \tag{5.59}$$

and (2) the impulse-modulated signal represented by a modulated delta-function series

$$\mathbf{u}(t; t_k) = \sum_{k=0}^{\infty} \mathbf{u}(t)\delta(t - t_k) \tag{5.60}$$

Signals having the forms of (5.59) and (5.60) are respectively depicted in Figs. 5.1a and 5.1b. In generating $\mathbf{u}(t_k)$ a zero order holding circuit of sorts is implied. The various t_k are arbitrary but satisfy the condition $t_{k+1} > t_k$.

From previous discussions the response of a continuous-time system to sampled inputs can be readily determined. In fact, it can be determined with considerably more ease than would be required to determine the response for all time. The continuous-time system is governed by the dynamic equation

$$\dot{\mathbf{x}} = \mathbf{A}\mathbf{x} + \mathbf{B}\mathbf{u}$$

wherein the solution is

$$\mathbf{x}(t) = \Phi(t)\mathbf{x}(t_0) + \int_{t_0}^{t} \Phi(t - \xi)\mathbf{B}\mathbf{u}(\xi)d\xi$$

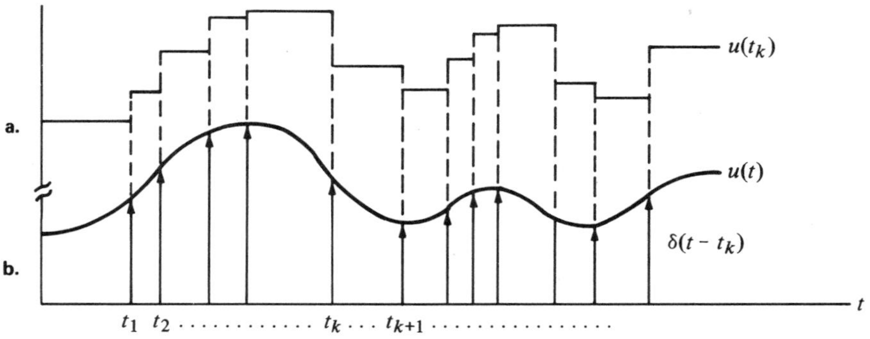

Figure 5.1 Sampled signal.

Let the input be the sampled signal of (5.59). It is a signal of constant amplitude over a prescribed time interval. Then

$$\mathbf{x}(t) = \Phi(t)\mathbf{x}(t_0) + \int_{t_0}^{t} \Phi(t-\xi)\mathbf{Bu}(t_k)d\xi \tag{5.61}$$

Now let the input consist of a *series* of constant vectors of form (5.59). Then at time $t_k + \tau$, where $0 < \tau \leqslant t_{k+1} - t_k$ for each time interval (t_k, t_{k+1}), we have

$$\mathbf{x}(t_k+\tau) = \Phi(\tau)\mathbf{x}(t_k) + \int_{t_k}^{t_k+\tau} \Phi(t_k+\tau-\xi)\mathbf{Bu}(t_k)d\xi \tag{5.62}$$

If the "staircase" input changes uniformly at intervals $t_k = kT$ equation (5.62) becomes

$$\mathbf{x}(kT+\tau) = \Phi(\tau)\mathbf{x}(kT) + \int_{kT}^{kT+\tau} \Phi(kT+\tau-\xi)\mathbf{Bu}(kT)d\xi \tag{5.63}$$

Finally, for $\tau = T$

$$\mathbf{x}(kT+T) = \Phi(T)\mathbf{x}(kT) + \int_{kT}^{kT+T} \Phi(kT+T-\xi)\mathbf{Bu}(kT)d\xi \tag{5.64}$$

Clearly (5.64) can be written as the difference equation

$$\mathbf{x}(k+1) = \bar{\mathbf{A}}\mathbf{x}(k) + \bar{\mathbf{B}}\mathbf{u}(k) \tag{5.65}$$

where

$$\bar{\mathbf{A}} = \Phi(T) = e^{\mathbf{A}t} \tag{5.66}$$

$$\bar{\mathbf{B}} = \int_0^T \Phi(T-\xi)d\xi \tag{5.67}$$

Thus, the continuous-time system governed by

$$\dot{\mathbf{x}} = \mathbf{Ax} + \mathbf{Bu}$$

having as the input a piece-wise constant signal is equivalent at sampling instants to the system governed by the difference equation (5.65).

To illustrate some of the theory developed consider a *simple* system governed by the differential equation

$$\left(\frac{d^2}{dt^2} + 5\frac{d}{dt} + 4\right)y(t) = u(t)$$

Let the input to the system be a continuous-time signal in one case, and in another case let it be the sampled signal $u(t) = 2^k$ for $kT < t \leqslant (k+1)T$, where $k = 0,1,2,\ldots,\infty$. From equation (4.19) the system matrices are

$$\mathbf{A} = \begin{bmatrix} 0 & 1 \\ -4 & -5 \end{bmatrix} \qquad \mathbf{B} = \begin{bmatrix} 0 \\ 1 \end{bmatrix}$$

$$\mathbf{C} = [1] \qquad \mathbf{D} = 0$$

The state transition matrix $\Phi(t)$ is the inverse Laplace Transform of $(s\mathbf{I} - \mathbf{A})^{-1}$, where

$$(s\mathbf{I} - \mathbf{A})^{-1} = \begin{bmatrix} s & 1 \\ -4 & s+5 \end{bmatrix}^{-1}$$

$$= \begin{bmatrix} s+5 & 1 \\ -4 & s \end{bmatrix} \frac{1}{(s+4)(s+1)}$$

By (5.39) and (5.44) we can evaluate the inverse transform of each polynomial element of the above matrix giving

$$\Phi(t) = \mathcal{L}^{-1}\{(s\mathbf{I} - \mathbf{A})^{-1}\}$$

$$= \frac{1}{3}\begin{bmatrix} 4e^{-t} - e^{-4t} & e^{-t} - e^{-4t} \\ -4(e^{-t} - e^{-4t}) & -e^{-t} + 4e^{-4t} \end{bmatrix}$$

For the case of a continuous-time input the state vector is, from (5.10),

$$\mathbf{x}(t) = \frac{1}{3}\begin{bmatrix} 4e^{-t} - e^{-4t} & e^{-t} - e^{-4t} \\ -4(e^{-t} - e^{-4t}) & -e^{-t} + 4e^{-4t} \end{bmatrix}\mathbf{x}(t_0)$$

$$+ \frac{1}{3}\int_{t_0}^{t} \begin{bmatrix} e^{-(t-\xi)} - e^{-4(t-\xi)} \\ -e^{-(t-\xi)} + 4e^{-4(t-\xi)} \end{bmatrix} u(\xi)d\xi$$

For the staircase input $u(t) = 2^k$ equation (5.64) specifies the state vector as the difference equation

$$\mathbf{x}(kT+T) = \frac{1}{3}\begin{bmatrix} 4e^{-T} - e^{-4T} & e^{-T} - e^{-4T} \\ -4(e^{-T} - e^{-4T}) & -e^{-T} + 4e^{-4T} \end{bmatrix} \mathbf{x}(kT)$$

$$+ \frac{1}{3}\begin{bmatrix} \frac{3}{4} - e^{-T} + \frac{1}{4}e^{-4T} \\ e^{-T} + e^{-4T} \end{bmatrix} 2^k$$

which is of the form

$$\mathbf{x}(k+1) = \bar{\mathbf{A}}\mathbf{x}(k) + \mathbf{B}\mathbf{u}(k)$$

Hence, by (5.56) the state vector for the continuous-time system having a staircase input is

$$\mathbf{x}(k) = \bar{\mathbf{A}}^k\mathbf{x}(0) + \sum_{\ell=0}^{k-1} \bar{\mathbf{A}}^{\ell}\bar{\mathbf{B}}\, 2^{k-\ell-1}$$

where $\bar{\mathbf{A}}$ and $\bar{\mathbf{B}}$ are determined by equations (5.66) and (5.67), respectively.

We next consider the continuous-time system wherein the input is the series of impulses represented by (5.60). Referring once again to state equation of a continuous-time system we have

$$\mathbf{x}(t) = \Phi(t)\mathbf{x}(t_0) + \int_{t_0}^{t} \Phi(t-\xi)\mathbf{B}\mathbf{u}(\xi)d\xi$$

At time $t_k + \tau$, where $0 < \tau \leqslant t_{k+1} - t_k$, we have on substituting (5.60) into the above state equation

$$\mathbf{x}(t_k+\tau) = \Phi(\tau)\mathbf{x}(t_k) + \Phi(\tau)\mathbf{B}\mathbf{u}(t_k) \tag{5.68}$$

Equation (5.68) is the state equation of a continuous-time system where the input is a series of impulses occurring at times t_k ($k = 0,1,2,\ldots\infty$). If the impulses occur at uniformly spaced time intervals where $t_{k+1} - t_k = T$, equation (5.68) gives the state equation for this condition as

$$\mathbf{x}(kT+\tau) = \Phi(\tau)\mathbf{x}(kT) + \Phi(\tau)\mathbf{B}\mathbf{u}(kT) \tag{5.69}$$

where $0 < \tau \leqslant T$. Lastly, for $\tau = T$ we have as the discrete-time state equation for a continuous-time system

$$\mathbf{x}(kT+T) = \Phi(T)\mathbf{x}(kT) + \Phi(T)\mathbf{Bu}(kT) \tag{5.70}$$

5.5 FIXED DISCRETE-TIME SYSTEMS: *z* DOMAIN ANALYSIS

Sampled data systems can be considered as continuous-time systems operating on discrete-time functions. The $\mathcal{Z}$ transform addressing discrete-time systems is introduced for the same reason that makes the Laplace transform useful in the study of continuous-time systems. Procedures similar to analyzing continuous-time systems in the frequency domain will be followed to analyze discrete-time systems in the z domain (where z is a complex variable).

The $\mathcal{Z}$ transform of a discrete-time function $f(k)$ is a power series in z^{-1}. The coefficients $f(k)$ are the amplitudes of the discrete-time signal. We have

$$\mathcal{Z}\{f(k)\} = \sum_{k=-\infty}^{\infty} f(k)z^{-k} \tag{5.71}$$

where $\mathcal{Z}\{f(k)\}$ is the $\mathcal{Z}$ transform of $f(k)$. For discrete-time systems those values of $f(k)$ where $k < 0$ are of less interest than where $k > 0$. The discussion to follow will, therefore, center around situations of positive values of k. Accordingly, our consideration of (5.71) will be bound by the values $k = 0,1,2,\ldots,\infty$.

From Appendix D the $\mathcal{Z}$ transform of the first forward difference equation is

$$\begin{aligned}\mathcal{Z}\{\Delta f(k)\} &= \mathcal{Z}\{f(k+1) - f(k)\}\\ &= (z-1)\mathcal{Z}\{f(k)\} - zf(0)\end{aligned} \tag{5.72}$$

where

$$\mathcal{Z}\{f(k)\} = \sum_{k=0}^{\infty} f(k)z^{-k} = f(0) + f(1)z^{-1} + f(2)z^{-2} + \ldots$$

$$\mathcal{Z}\{f(k+1)\} = f(1) + f(2)z^{-1} + f(3)z^{-2} + \ldots = z[\mathcal{Z}\{f(k)\} + f(0)] \tag{5.73}$$

The characterization of a linear discrete-time system of fixed form was specified earlier by equations (5.50) and (5.51):

$$\mathbf{x}(k+1) = \mathbf{A}\mathbf{x}(k) + \mathbf{B}\mathbf{u}(k)$$

$$\mathbf{y}(k) = \mathbf{C}\mathbf{x}(k) + \mathbf{D}\mathbf{u}(k)$$

Taking the $\mathcal{Z}$ transform of both sides of the above normal equations we have, using property (5.73),

$$z\mathbf{X}(z) - z\mathbf{x}(0) = \mathbf{A}\mathbf{X}(z) + \mathbf{B}\mathbf{U}(z) \tag{5.74}$$

$$\mathbf{Y}(z) = \mathbf{C}\mathbf{X}(z) + \mathbf{D}\mathbf{U}(z) \tag{5.75}$$

where

$$\mathbf{U}(z) = \mathcal{Z}\{\mathbf{u}(k)\}$$

$$\mathbf{X}(z) = \mathcal{Z}\{\mathbf{x}(k)\}$$

$$\mathbf{Y}(z) = \mathcal{Z}\{\mathbf{y}(k)\}$$

Solving (5.74) for $\mathbf{X}(z)$ we have

$$(z\mathbf{I} - \mathbf{A})\mathbf{X}(z) = z\mathbf{x}(0) + \mathbf{B}\mathbf{U}(z)$$

$$\mathbf{X}(z) = (z\mathbf{I} - \mathbf{A})^{-1}z\mathbf{x}(0) + (z\mathbf{I} - \mathbf{A})^{-1}\mathbf{B}\mathbf{U}(z) \tag{5.76}$$

which, when substituted in (5.75), gives the discrete-time output as

$$\mathbf{Y}(z) = \mathbf{C}(z\mathbf{I} - \mathbf{A})^{-1}z\mathbf{x}(0) + [\mathbf{C}(z\mathbf{I} - \mathbf{A})^{-1}\mathbf{B} + \mathbf{D}]\mathbf{X}(z) \tag{5.77}$$

Referring to state equations (5.56) and (5.58) the $\mathcal{Z}$ transform of the output equation (5.58) gives

$$\mathbf{Y}(z) = \mathbf{C}\hat{\Phi}(z)\mathbf{x}(0) + [\mathbf{C}\hat{\Phi}(z)z^{-1}\mathbf{B} + \mathbf{D}]\mathbf{X}(z) \tag{5.78}$$

where

$$\hat{\Phi}(z) = \mathcal{Z}\{\Phi(k)\} = \mathcal{Z}\{\mathbf{A}^k\} \tag{5.79}$$

Comparing (5.78) with (5.77) we see that

$$\hat{\Phi}(z) = (z\mathbf{I} - \mathbf{A})^{-1}z = (\mathbf{I} - \mathbf{A}z^{-1})^{-1} \tag{5.80}$$

Result (5.80) can also be derived directly from the definition of the $\mathcal{Z}$ transform for $\mathbf{A}^k$:

$$\mathcal{Z}\{\mathbf{A}^k\} = \sum_{k=0}^{\infty} \mathbf{A}^k z^{-k}$$

$$= (z\mathbf{I} - \mathbf{A})^{-1} z \tag{5.81}$$

The zero-state response specified by (5.78) is

$$\mathbf{Y}(z) = [\mathbf{C}\hat{\Phi}(z)z^{-1}\mathbf{B} + \mathbf{D}]\mathbf{X}(z) \tag{5.82}$$

Hence, the transfer function matrix for a discrete-time system is

$$\mathbf{H} = \mathbf{C}\hat{\Phi}(z)z^{-1}\mathbf{B} + \mathbf{D} \tag{5.83}$$

Proceeding in the same manner as in Section 5.2 the inverse transform operator $\hat{\Phi}(z)$ can be written as

$$\hat{\Phi}(z) = z(z\mathbf{I} - \mathbf{A})^{-1} = \frac{z \operatorname{adj}(z\mathbf{I} - \mathbf{A})}{|z\mathbf{I} - \mathbf{A}|}$$

$$= \frac{z\mathbf{B}(z)}{|z\mathbf{I} - \mathbf{A}|} = \frac{z\mathbf{B}(z)}{p(z)} \tag{5.84}$$

where $\mathbf{B}(z)$ and $p(z)$ are polynomials in z. Replacing s with z in (5.26) and (5.32) the ratio $\hat{\Phi}(z)$ can be evaluated as prescribed by (5.31) through (5.35). Representing $\mathbf{B}(z)$ and $p(z)$, respectively, as

$$\mathbf{B}(z) = \mathbf{B}_0 z^{k-1} + \mathbf{B}_1 z^{k-2} + \ldots + \mathbf{B}_{k-2} z + \mathbf{B}_{k-1}$$

and

$$p(z) = z^k + a_1 z^{k-1} + a_2 z^{k-2} + \ldots + a_{k-1} z + a_k$$

equation (5.84) becomes

$$\hat{\Phi}(z) = \frac{\mathbf{B}_0 z^{k-1} + \mathbf{B}_1 z^{k-2} + \ldots + \mathbf{B}_{k-2} z + \mathbf{B}_{k-1}}{z^{k-1} + a_1 z^{k-2} + \ldots + a_{k-1} + a_k z^{-1}} \tag{5.85}$$

Clearing the fraction we have

$$(z^k + a_1 z^{k-1} + a_2 z^{k-2} + \ldots + a_1 z + a_k)\mathbf{I}$$

$$= (\mathbf{B}_0 z^{k-1} + \mathbf{B}_1 z^{k-2} + \ldots + \mathbf{B}_{k-2} z + \mathbf{B}_{k-1})(z\mathbf{I} - \mathbf{A})$$

$$= \mathbf{B}_0 z^k - \mathbf{B}_0 \mathbf{A} z^{k-1} + \mathbf{B}_1 z^{k-1} + \ldots + \mathbf{B}_{k-1} z - \mathbf{B}_{k-1}\mathbf{A}$$

Equating the coefficients of like powers of z gives

$$\begin{aligned}
\mathbf{B}_0 &= \mathbf{I} & a_1 &= -\operatorname{tr}\mathbf{A} \\
\mathbf{B}_1 &= \mathbf{B}_0\mathbf{A} + a_1\mathbf{I} & a_2 &= -\frac{1}{2}\operatorname{tr}(\mathbf{B}_1\mathbf{A}) \\
\mathbf{B}_2 &= \mathbf{B}_1\mathbf{A} + a_2\mathbf{I} & a_3 &= -\frac{1}{3}\operatorname{tr}(\mathbf{B}_2\mathbf{A}) \\
&\cdots\cdots\cdots & &\cdots\cdots\cdots \\
\mathbf{B}_{k-1} &= \mathbf{B}_{k-2}\mathbf{A} + a_{k-1}\mathbf{I} & a_{k-1} &= -\frac{1}{k-1}\operatorname{tr}(\mathbf{B}_{k-2}\mathbf{A}) \\
0 &= \mathbf{B}_{k-1}\mathbf{A} + a_k\mathbf{I} & a_k &= -\frac{1}{k}\operatorname{tr}(\mathbf{B}_{k-1}\mathbf{A})
\end{aligned} \tag{5.86}$$

which results in the algorithms

$$a_i = -\frac{1}{i}\operatorname{tr}(\mathbf{B}_{i-1}\mathbf{A}) \qquad (i = 1,2,\ldots,k) \tag{5.87}$$

$$\mathbf{B}_i = \mathbf{B}_{i-1}\mathbf{A} + a_i\mathbf{I} \qquad (i = 1,2,\ldots,k-1) \tag{5.88}$$

The most commonly used methods to determine $\Phi(k)$, the inverse transform of (5.84), include inversion by partial fractions, inversion using the inversion integral, or inversion by long division. Each of these techniques requires addressing the individual elements of the matrix $\hat{\Phi}(z)$. Consequently, the techniques are laborious. To determine the signal element $f(k)$ corresponding to each element $f(z)$ in $\hat{\Phi}(z)$ it becomes necessary to expand $f(z)$ into a power series in z^{-1}. The expansion coefficients $f(k)$ are the amplitudes of the discrete-time signal.

Inversion by partial fractions involves factoring and looking-up in an appropriate table the discrete-time function corresponding to the transform to be inverted. Appendix D contains a typical conversion table, which of course, can be expanded considerably. The task of factoring the ratio of two polynomials was discussed earlier. The associated z-formulation takes the form of equations (5.36) through (5.40).

A more elegant technique, which also involves factoring, is use of the inversion integral

$$f(k) = \frac{1}{2\pi j} \oint_c f(z) z^{k-1} dz \tag{5.89}$$

From the calculus of residues the evaluation of (5.89) results in

$$f(k) = \sum \text{residues of } f(z)z^{k-1} \text{ at its poles within circle of convergence of } f(z). \tag{5.90}$$

For a pole at $z = \lambda_j$ of multiplicity α_j the residue of $f(z)z^{k-1}$ is given by

$$\left.\begin{matrix}\text{Residue } f(z)z^{k-1} \text{ at} \\ \text{pole } z = \lambda_j, \text{ order } \alpha_j\end{matrix}\right\} = \lim_{z \to \lambda_j} \frac{1}{(\alpha_j - 1)!} \frac{d^{\alpha_j - 1}}{dz^{\alpha_j - 1}} (z - \lambda_j)^{\alpha_j} f(z) z^{k-1} \tag{5.91}$$

A third method of inversion of the $\mathcal{Z}$ transform applies when $f(z)$ is a rational function. Both the numerator and denominator can be expressed as polynomials in z^{-1}. Using long division the numerator is divided by the denominator giving a series expression in powers of z^{-1}. The numerical values of the coefficients are the $f(k)$. This method is particularly useful when tables of $\mathcal{Z}$ transforms or the inversion integral cannot be used. However, a general expression for $f(k)$, as can be obtained by the other two methods, is not readily achieved through the long division method.

To illustrate the above three inversion methods consider the following example. We want to determine the inverse transform of the simple complex function

$$f(z) = \frac{2z^2 - 4z}{z^2 - 4z + 3}$$

by each of the three methods.

(1) *Partial Fractions.* Factoring $f(z)$ we can write

$$f(z) = \frac{2z^2 - 4z}{z^2 - 4z + 3}$$

$$= \frac{2z(z-2)}{(z-1)(z-3)}$$

$$= \frac{z}{z-1} + \frac{z}{z-3}$$

From Appendix D:

$$f(k) = 1 + 3^k$$

(2) *Inversion Formula.* The function $f(z)z^{k-1}$ has simple poles at $z = 1$ and $z = 3$. There is no pole at the origin. Thus from (5.90)

$$f(k) = \left.\frac{2z^2 - 4z}{z-3} z^{k-1}\right|_{z=1} + \left.\frac{2z^2 - 4z}{z-1} z^{k-1}\right|_{z=3}$$

$$= 1 + 3^k$$

(3) *Long Division.* Expressing both the numerator and denominator of $f(z)$ as polynomials in z^{-1} we have

$$f(z) = \frac{2 - 4z^{-1}}{1 - 4z^{-1} + 3z^{-2}}$$

The long division gives

$$\begin{array}{r l} & 2 + 4z^{-1} + 10z^{-2} + 28z^{-3} + 82z^{-4} + \ldots \\ \hline 1 - 4z^{-1} + 3z^{-2}\,) & 2 - 4z^{-1} \\ & \underline{2 - 8z^{-1} + 6z^{-2}} \\ & \ldots\ldots\ldots\ldots\ldots \end{array}$$

$$f(0) = 2 = 1 + 3^0$$

$$f(1) = 4 = 1 + 3$$

$$f(2) = 10 = 1 + 3^2$$

$$\ldots\ldots\ldots\ldots\ldots$$

$$f(k) = 1 + 3^k$$

5.6 FINITE-STATE MARKOV SYSTEMS: INPUT-FREE

The analysis considered thus far has been *deterministic*[1] in nature, i.e., it was assumed that the system states at time t could be determined with certainty. This certainty was a manifestation of exact knowledge of the characterization matrices, the state at some initial time t_0, and the input applied over the time interval (t_0,t). For those systems, such as large scale systems, where the characterization matrices are not known exactly, it can be expected that the system state and output can only be described statistically—even though the input and initial state are known exactly. These systems are called *stochastic systems*. They are characterized in terms of *probabilities* of being in a specific state at some given time, and in terms of *state transition* probabilities associated with each discrete state. If the discrete states that the system can assume are finite in number then such systems are *finite-state stochastic systems*.

In formulating the probabilistic description of finite-state stochastic systems we define the following quantities: Let the probability that the system is in discrete state j at time $(k + 1)$ be $p_j(k + 1)$. At an earlier time k the system can be in any discrete state i with probability $p_i(k)$. For each state i there corresponds a state transition probability φ_{ji} that the system will transition from state i to state j. We stipulate that φ_{ji} is independent of the past history of the system, i.e., it is independent of how the system arrived at state j from state i. (Systems that possess this property are known as Markov systems.[2]) Thus, the probability that the system will be in state j at time $(k + 1)$ is

$$p_j(k+1) = \sum_{i=1}^{n} \varphi_{ji}(k)p_i(k) \qquad j = 1,2,\ldots,n \tag{5.92}$$

where n is the total number of discrete states the system can assume. (The index n should not be confused with our previous n used in establishing the order of the differential equation governing the continuous-time system.) Equation (5.92) can be written in vector form as

$$\mathbf{p}(k+1) = \Phi(k)\mathbf{p}(k) \tag{5.93}$$

1. For the sake of completeness our previous formulation assumed that all defined quantities were readily calculable. It must be recognized that this is true only for simple cases. For complex systems these quantities are not readily ascertained and must be treated statistically.
2. For a more complete discussion of Markov chains see, for example, Feller, W., An Introduction to Probability Theory and Its Applications, Vol. I, 3rd Edition, John Wiley & Sons, New York (1968).

where $\mathbf{p}$ is an n-dimensional *state-probability* vector and Φ is an $n \times n$ *state transition* probability matrix. Clearly, from (5.92) and (5.93) Φ is of the form

$$\Phi(k) = \begin{bmatrix} \varphi_{11}(k) & \varphi_{12}(k) & \dots & \varphi_{1n}(k) \\ \dots & \dots & \dots & \dots \\ \varphi_{n1}(k) & \varphi_{n2}(k) & \dots & \varphi_{nn}(k) \end{bmatrix} \tag{5.94}$$

where φ_{ji} must satisfy the following probability conditions:

$$\varphi_{ji} \geqslant 0 \qquad \text{for all } i,j \tag{5.95}$$

$$\sum_{j=1} \varphi_{ji} = 1 \qquad \text{for all } i \tag{5.96}$$

If the elements φ_{ji} are fixed then (5.93) becomes

$$\mathbf{p}(k+1) = \Phi \mathbf{p}(k) \tag{5.97}$$

The derivation of (5.97) was based solely on probabilistic reasoning; independent of the linear, time-invariant conditions of Chapter 3. However, formula (5.97) which describes a zero-input probabilistic system, is of the same form as state equation (5.50), which describes a zero-input, linear, time-invariant system:

$$\mathbf{x}(k+1) = \Phi \mathbf{x}(k) \tag{5.50}$$

Moreover, the system described by (5.97) need only be a finite-state Markov system. The likeness of (5.50) and (5.96) suggests that a class of nonlinear systems, where the states are represented in terms of probabilities, is governed by linear equations. Therefore, the solution of (5.97) could be obtained by following the methods of Sections 3 and 5. By direct recursion (5.97) becomes

$$\mathbf{p}(k) = \Phi^k \mathbf{p}(0) \tag{5.98}$$

Hence, for input-free Markov systems the state probabilities are completely "determined" by knowing the transition probability matrix and the initial state probability vector. The evaluation of (5.98) is best carried-out in the z domain, wherein the $\mathcal{Z}$ transform gives

$$z\mathbf{P}(z) - z\mathbf{p}(0) = \Phi \mathbf{P}(z)$$

$$\mathbf{P}(z) = (\mathbf{I} - \Phi z^{-1})^{-1} \mathbf{p}(0) \tag{5.99}$$

We demonstrate the use of (5.99) through simple examples below.

In general Markov systems can be represented graphically by state diagrams (signal flow graphs) as shown in Fig. 5.2. The nodes are the states i and the directed branches are the state transition probabilities φ_{ji}. The states can be grouped into mutually exclusive state sets, whereby the system can change from every state to every other state within each state set. Those state sets that once entered can never be left are called *ergodic sets*. Their associated states are *recurrent states*. Those state sets that once left can never again be entered are called *transient sets*. Generally, Markov systems can have many ergodic and/or transient state sets as shown in Fig. 5.3. States 1, 2, and 5, 6 comprise ergodic sets whereas states 3, 4 are a transient set. Each set is mutually exclusive of the other. As a minimum a Markov system must have at least one ergodic set. Such a system is an ergodic system. For a system having more than one ergodic set but no transient sets the state sets are said to be disjoint. The system can be separated into a number of independent subsystems, one for each ergodic set. Further, if an ergodic set has at least one state such that the system must return to this state periodically the ergodic set is *periodic*, otherwise it is regular. We shall expand the above statements through the examples below.

Consider the simple system of Fig. 5.4, which has an ergodic set but no transient set. We desire to evaluate the probability that this system will transition to states 2 or 3, given that the system is in state 1 at time $k = 0$. The transition matrix for the system of Fig. 5.4 is

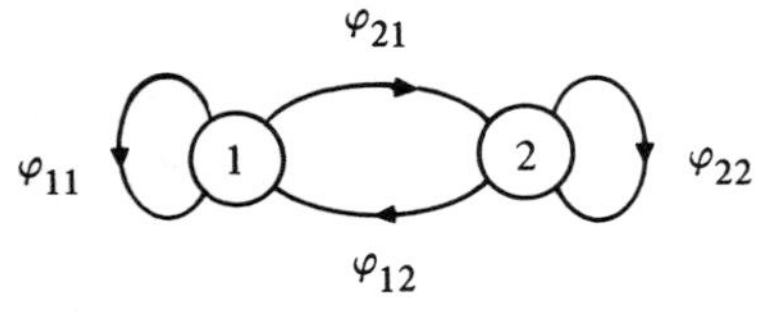

Figure 5.2 Two-state Markov system.

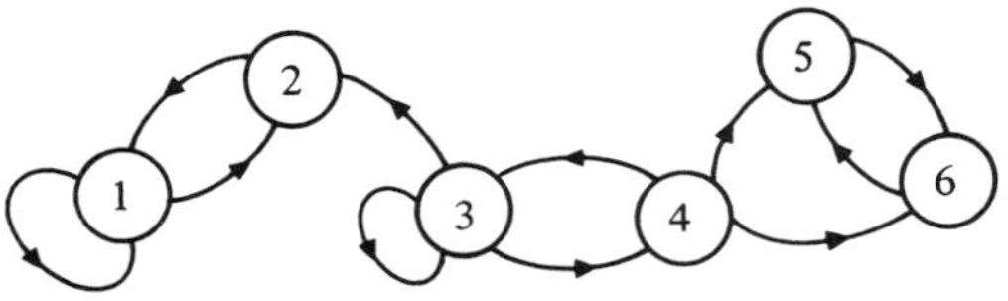

Figure 5.3 Six-state Markov system.

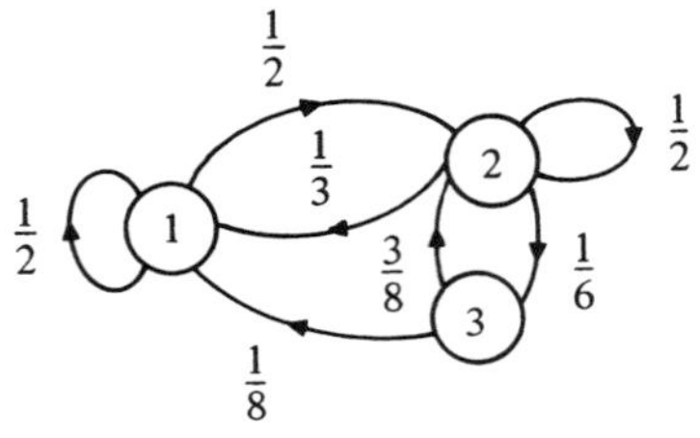

Figure 5.4 Simple ergodic system.

$$\mathbf{\Phi} = \begin{bmatrix} \frac{1}{2} & \frac{1}{3} & \frac{1}{8} \\ \frac{1}{2} & \frac{1}{2} & \frac{3}{8} \\ 0 & \frac{1}{6} & \frac{1}{2} \end{bmatrix}$$

where the transition probabilities satisfy conditions (5.95) and (5.96). Given that the system is in state 1 at time $k = 0$ the initial state probability vector is

$$\mathbf{p}(0) = \begin{bmatrix} 1 \\ 0 \\ 0 \end{bmatrix}$$

From (5.99)

$$\begin{aligned} \mathbf{P}(z) &= (\mathbf{I} - \mathbf{\Phi} z^{-1})^{-1}\mathbf{p}(0) \\ &= \begin{bmatrix} 1 - \frac{1}{2}z^{-1} & -\frac{1}{3}z^{-1} & -\frac{1}{8}z^{-1} \\ -\frac{1}{2}z^{-1} & 1 - \frac{1}{2}z^{-1} & -\frac{3}{8}z^{-1} \\ 0 & -\frac{1}{6}z^{-1} & 1 - \frac{1}{2}z^{-1} \end{bmatrix}^{-1} \begin{bmatrix} 1 \\ 0 \\ 0 \end{bmatrix} \end{aligned}$$

Evaluating the inverse matrix by cofactors we have

$$\mathbf{P}(z) = \frac{1}{|\mathbf{I} - \mathbf{\Phi} z^{-1}|} \begin{bmatrix} 1 - z^{-1} + \frac{3}{16}z^{-2} & \frac{1}{3}z^{-1} - \frac{7}{48}z^{-2} & \frac{1}{8} - \frac{1}{16}z^{-1} + \frac{1}{8}z^{-2} \\ \frac{1}{2}z^{-1} - \frac{1}{4}z^{-2} & 1 - z^{-1} + \frac{1}{4}z^{-2} & \frac{3}{8}z^{-1} - \frac{1}{8}z^{-2} \\ \frac{1}{12}z^{-1} & \frac{1}{6}z^{-1} - \frac{1}{12}z^{-2} & 1 - z^{-1} + \frac{1}{12}z^{-2} \end{bmatrix} \begin{bmatrix} 1 \\ 0 \\ 0 \end{bmatrix}$$

where the determinant

$$|\mathbf{I} - \mathbf{\Phi} z^{-1}| = 1 - \frac{25}{16}z^{-1} + \frac{7}{12}z^{-2} - \frac{1}{48}z^{-3}$$

Hence,

$$\mathbf{P}(z) = \begin{bmatrix} P_1(z) \\ P_2(z) \\ P_3(z) \end{bmatrix} = \frac{1}{|\mathbf{I} - \Phi z^{-1}|} \begin{bmatrix} 1 - z^{-1} + \frac{3}{16} z^{-2} \\ \frac{1}{2} z^{-1} - \frac{1}{4} z^{-2} \\ \frac{1}{12} z^{-2} \end{bmatrix}$$

As was mentioned earlier there exists for small systems a variety of ways in which the above matrix can be evaluated. For discussion purposes we shall employ the methods of long division and factoring. The method of long division gives

$$\mathbf{P}(z) = \begin{bmatrix} 1 + .5625z^{-1} + .4831z^{-2} + .4379z^{-3} + .4141z^{-4} + .4074z^{-5} + \ldots \\ 0 + .5z^{-1} + .5381z^{-2} + .5492z^{-3} + .5546z^{-4} + .5574z^{-5} + \ldots \\ 0 + 0 + .0833z^{-2} + .1302z^{-3} + .1548z^{-4} + .1676z^{-5} + \ldots \end{bmatrix}$$

where the decimal values are rounded off at the fourth place. From (5.73) the coefficients of z^{-k} give the probability amplitudes $p_i(k)$ directly; thus we have the discrete-time histories of the three components of $\mathbf{p}(k)$:

$p_1(0) = 1$	$p_2(0) = 0$	$p_3(0) = 0$
$p_1(1) = .5628$	$p_2(1) = .5$	$p_3(1) = 0$
$p_1(2) = .4831$	$p_2(2) = .5381$	$p_3(2) = .0833$
$p_1(3) = .4379$	$p_2(3) = .5492$	$p_3(3) = .1302$
$p_1(4) = .4141$	$p_2(4) = .5546$	$p_3(4) = .1548$
$p_1(5) = .4074$	$p_2(5) = .5574$	$p_3(5) = .1676$
....	....	....

The above state probabilities as a function of discrete-time are shown graphically in Fig. 5.5.

We observe that as k becomes larger the probability of the system remaining in state 1 decreases sharply and approaches a probability level of 0.4. Knowledge of the initial state ($k = 0$) becomes less significant with time. The probability at time $k \gg 0$ of the system being in a state other than state 1 is 0.6. In fact

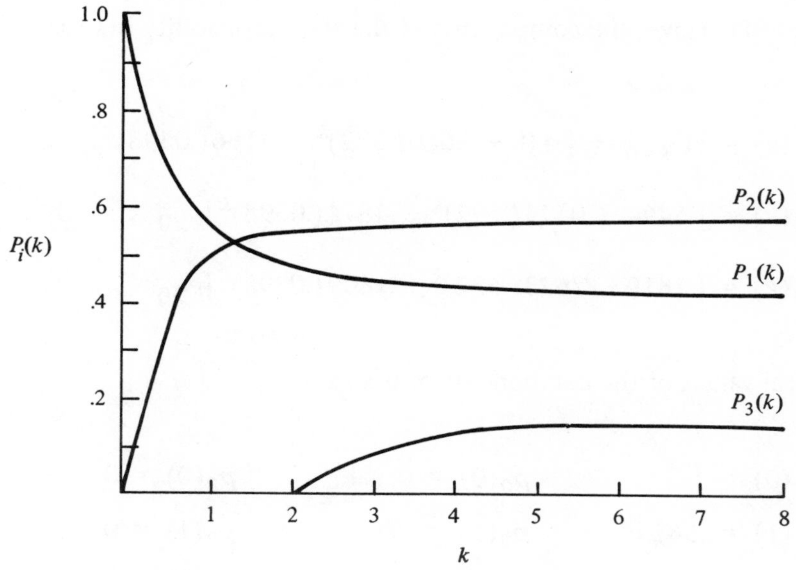

Figure 5.5

the system is in state 2 with probability greater than 0.5 and in state 3 with probability 0.16. (The fact that the total probability exceeds unity is attributed to rounding-off of decimal quantities in the calculations.)

As an alternative method for evaluating the matrix $\mathbf{P}(z)$ of our example system we employ the method of partial fractions. The system determinant can be factored as

$$|\mathbf{I} - \Phi z^{-1}| = 1 - \frac{25}{16} z^{-1} + \frac{7}{12} z^{-2} - \frac{1}{48} z^{-3}$$

$$= z^{-3}(z - 1)(z - .522)(z - .0398)$$

thus

$$\mathbf{P}(z) = \frac{z^3}{(z - 1)(z - .522)(z - .0398)} \begin{bmatrix} 1 - z^{-1} + \frac{3}{16} z^{-2} \\ \frac{1}{2} z^{-1} - \frac{1}{4} z^{-2} \\ \frac{1}{12} z^{-2} \end{bmatrix}$$

The inverse of $\mathbf{P}(z)$ gives the components of the state probability $\mathbf{p}(k)$ as

$$p_1(k) = (1)_{k=0} + [.410 + .2680(.522)^k + .3166(.0398)^k]_{k>0}$$

$$p_2(k) = [.5494 - .0574(.522)^k - .4874(.0398)^k]_{k>0}$$

$$p_3(k) = [.1819 - .3623(.522)^k - .1809(.0398)^k]_{k>0}$$

A few typical values of the components of $\mathbf{p}(k)$ are

$p_1(0) = 1$	$p_2(0) = 0$	$p_3(0) = 0$
$p_1(1) = .5625$	$p_2(1) = .5$	$p_3(1) = 0$
$p_1(2) = .4835$	$p_2(2) = .5331$	$p_3(2) = .0835$
$p_1(3) = .4481$	$p_2(3) = .5413$	$p_3(3) = .1314$
$p_1(4) = .4299$	$p_2(4) = .5452$	$p_3(4) = .1551$
....	....	....
$p_1(\infty) = .410$	$p_2(\infty) = .5494$	$p_3(\infty) = .1819$

which compare favorably with the results achieved by the method of long division.

Analysis by partial fractions provides for system insight not readily attainable through analysis by long division. We have seen by both methods that the components of the state probability vector $\mathbf{p}(k)$ must not exceed unity. Since $|p_i(k)| \leqslant 1$ for all i and k it follows from the method of partial fractions that none of the eigenvalues of $\mathbf{\Phi}$ can have a magnitude greater than unity. Thus all roots of magnitude less than unity contribute transient terms which vanish as $k \to \infty$. Since every Markov system must have at least one ergodic set it also follows that every Markov system must have at least one root equal to unity. In fact the characteristic equation

$$|\mathbf{I} - \mathbf{\Phi} z^{-1}| = 0$$

of a Markov system will have as many roots equal to unity as there are ergodic sets in the system.

5.7 TIME-VARYING CONTINUOUS-TIME SYSTEMS: TIME DOMAIN ANALYSIS

For time-varying linear systems the differential equations to be solved are

$$\dot{\mathbf{x}}(t) = \mathbf{A}(t)\mathbf{x}(t) + \mathbf{B}(t)\mathbf{u}(t)$$

$$\mathbf{y}(t) = \mathbf{C}(t)\mathbf{x}(t) + \mathbf{D}(t)\mathbf{u}(t)$$

The homogeneous form for the dynamical equation is

$$\frac{d}{dt}\mathbf{x}(t) = \mathbf{A}(t)\mathbf{x}(t) \tag{5.100}$$

We begin the solution to (5.100) by assuming that the components of $\mathbf{x}(t)$ are linearly related to the components of $\mathbf{x}(t_0)$. Using definition (3.35) this relationship for zero input is

$$x_i(t) = \sum_j \varphi_{ij}(t,t_0)x_j(t_0) \tag{5.101}$$

wherein the initial conditions are established as

$$x_i(t_0) = 0 \qquad \text{for } i \neq j$$

$$x_j(t_0) \neq 0$$

Thus (5.101) represents the solution for an arbitrary state by superposition of the initial conditions. In matrix form (5.101) becomes

$$\mathbf{x}(t) = \Phi(t,t_0)\mathbf{x}(t_0) \tag{5.102}$$

Returning to the dynamical equation (5.100), if it is of first order we can write in scalar form

$$\frac{dx}{x} = a(t)dt \tag{5.103}$$

Integrating both sides of (5.103) from t_0 to t gives

$$\ln \frac{x(t)}{x(t_0)} = \int_{t_0}^{t} a(\xi)\, d\xi$$

$$x(t) = x(t_0) \exp\left(\int_{t_0}^{t} a(\xi)\, d\xi \right) \tag{5.104}$$

By (5.104) it is inferred that for first order systems the fundamental matrix is

$$\varphi(t,t_0) = \exp\left(\int_{t_0}^{t} a(\xi)\, d\xi \right) \tag{5.105}$$

This implies that for the initial conditions we let

$$x_i(t) = \varphi_{ij}(t,t_0) x_j(t_0)$$

$$x_i(t_0) = 0 \qquad \text{for } i \neq j$$

$$x_j(t_0) = 1$$

Extending (5.105) to higher order systems we have

$$\Phi(t,t_0) = \exp\left(\int_{t_0}^{t} \mathbf{A}(\xi)\, d\xi \right) \tag{5.106}$$

Relation (5.106) is valid only if $\mathbf{A}(t)$ and the integral equation commute, i.e., if

$$\mathbf{A}(t) \int_{t_0}^{t} \mathbf{A}(\xi) d\xi - \int_{t_0}^{t} \mathbf{A}(\xi) \mathbf{A}(t) d\xi = 0 \tag{5.107}$$

Clearly, for the case where $\mathbf{A}(t)$ is the product of a constant matrix and a scalar time function, such as

$$\mathbf{A}(t) = f(t)\mathbf{A}'$$

where

$$\int_{t_0}^{t} \mathbf{A}(\xi)d\xi = \mathbf{A}' \int_{t_0}^{t} f(\xi)d\xi$$

then identity (5.107) holds. Thus, we let

$$\Phi(t,t_0) = \exp\left(\mathbf{A}' \int_{t_0}^{t} f(\xi)d\xi\right) \tag{5.108}$$

In general, however, it is not possible to obtain an analytic expression for the fundamental matrix of a time-varying linear system. It is primarily of conceptual interest.

To find the complete solution to the dynamical equation we again refer to the method of variation of the parameter. Proceeding in a manner analogous to that of the previous section the solution $\mathbf{x}(t)$ is assumed to have the form

$$\mathbf{x}(t) = \Phi(t,t_0)\mathbf{f}(t) \tag{5.109}$$

where $\mathbf{f}(t)$ is to be determined. The time derivative of (5.109) is

$$\dot{\mathbf{x}}(t) = \dot{\Phi}(t,t_0)\mathbf{f}(t) + \Phi(t,t_0)\dot{\mathbf{f}}(t)$$

which, when substituted into (2.26), gives

$$[\dot{\Phi}(t,t_0) - \mathbf{A}(t)\Phi(t,t_0)]\mathbf{f}(t) + \Phi(t,t_0)\dot{\mathbf{f}}(t) = \mathbf{B}(t)\mathbf{u}(t)$$

The expression in the bracket is the homogenous equation and is identically zero, leaving

$$\Phi(t,t_0)\dot{\mathbf{f}}(t) = \mathbf{B}(t)\mathbf{u}(t) \tag{5.110}$$

Multiplying both sides of (5.110) from the left by the inverse of $\Phi(t,t_0)$ and integrating we obtain the expression for $\mathbf{f}(t)$:

$$\mathbf{f}(t) = \int_{-\infty}^{t} \Phi^{-1}(\xi,t_0)\mathbf{B}(\xi)\mathbf{u}(\xi)d\xi$$

Thus, solution (5.109) can be written as

$$\mathbf{x}(t) = \int_{-\infty}^{t} \Phi(t,t_0)\Phi^{-1}(\xi,t_0)\mathbf{B}(\xi)\mathbf{u}(\xi)d\xi$$

$$= \Phi(t,t_0)\int_{-\infty}^{t_0} \Phi^{-1}(\xi,t_0)\mathbf{B}(\xi)\mathbf{u}(\xi)d\xi$$

$$+ \int_{t_0}^{t} \Phi(t,t_0)\Phi^{-1}(\xi,t_0)\mathbf{B}(\xi)\mathbf{u}(\xi)d\xi \qquad (5.111)$$

Evaluating the above equations at $t = t_0$ results in

$$\mathbf{x}(t_0) = \int_{-\infty}^{t_0} \Phi(t,t_0)\Phi^{-1}(\xi,t_0)\mathbf{B}(\xi)\mathbf{u}(\xi)d\xi$$

$$= \int_{-\infty}^{t_0} \Phi^{-1}(\xi,t_0)\mathbf{B}(\xi)\mathbf{u}(\xi)d\xi$$

where, from (5.3), $\Phi(t_0,t_0) = \mathbf{I}$. Further, it can readily be shown that

$$\Phi^{-1}(\xi,t_0) = \Phi(t_0,\xi)$$

Thus, solution (5.111) becomes

$$\mathbf{x}(t) = \Phi(t,t_0)\mathbf{x}(t_0) + \int_{t_0}^{t} \Phi(t,t_0)\Phi(t_0,\xi)\mathbf{B}(\xi)\mathbf{u}(\xi)d\xi$$

$$= \Phi(t,t_0)\mathbf{x}(t_0) + \int_{t_0}^{t} \Phi(t,\xi)\mathbf{B}(\xi)\mathbf{u}(\xi)d\xi \qquad (5.112)$$

From (2.27) the input-output-state-relationship for the time-varying system is

$$\mathbf{y}(t) = \mathbf{C}(t)\left[\Phi(t,t_0)\mathbf{x}(t_0) + \int_{t_0}^{t} \Phi(t,\xi)\mathbf{B}(\xi)\mathbf{u}(\xi)d\xi\right] + \mathbf{D}(t)\mathbf{u}(t) \qquad (5.113)$$

5.8 TIME-VARYING DISCRETE TIME SYSTEMS: TIME DOMAIN ANALYSIS

The normal form of the equations governing a linear time-varying discrete-time system is

$$\mathbf{x}(k+1) = \mathbf{A}(k)\mathbf{x}(k) + \mathbf{B}(k)\mathbf{u}(k) \tag{5.114}$$

$$\mathbf{y}(k) = \mathbf{C}(k)\mathbf{x}(k) + \mathbf{D}(k)\mathbf{u}(k) \tag{5.115}$$

As shown before the solution to (5.114) can be found directly by recursion:

$$\begin{aligned}
\mathbf{x}(1) &= \mathbf{A}(0)\mathbf{x}(0) + \mathbf{B}(0)\mathbf{u}(0)\\
\mathbf{x}(2) &= \mathbf{A}(1)\mathbf{x}(1) + \mathbf{B}(1)\mathbf{u}(1)\\
&= \mathbf{A}(1)\mathbf{A}(0)\mathbf{x}(0) + \mathbf{A}(1)\mathbf{B}(0)\mathbf{u}(0) + \mathbf{B}(1)\mathbf{u}(1)\\
&\cdots\cdots\cdots\cdots\cdots\cdots
\end{aligned} \tag{5.116}$$

For $k > \ell$ we define

$$\Phi(k,\ell) = \prod_{i=\ell}^{k-1} \mathbf{A}(i) = \mathbf{A}(k-1)\mathbf{A}(k-2)\ldots\mathbf{A}(\ell+1)\mathbf{A}(\ell) \tag{5.117}$$

and

$$\Phi(k,k) = \mathbf{I} \tag{5.118}$$

Thus the equation of state for a non-stationary, discrete-time system can be written as

$$\mathbf{x}(k) = \Phi(k,0)\mathbf{x}(0) + \sum_{\ell=0}^{k-1} \Phi(k,\ell+1)\mathbf{B}(\ell)\mathbf{u}(\ell) \tag{5.119}$$

Substituting (5.119) into (5.115) the output expression becomes

$$\mathbf{y}(k) = \mathbf{C}(k)\left[\Phi(k,0)\mathbf{x}(0) + \sum_{\ell=0}^{k-1} \Phi(k,\ell+1)\mathbf{B}(\ell)\mathbf{u}(\ell)\right] + \mathbf{D}(k)\mathbf{u}(k) \tag{5.120}$$

For zero initial conditions the system output is

$$\mathbf{y}(k) = \sum_{\ell=0}^{k-1} \mathbf{C}(k)\Phi(k,\ell+1)\mathbf{B}(\ell)\mathbf{u}(\ell) \tag{5.121}$$

From our previous definition of the transfer function matrix H, where

$$y = Hu$$

the weighting sequence matrix $\mathbf{H}$ for non-stationary systems has the form, from (5.121),

$$\mathbf{H}(k, k-\ell) = \mathbf{C}(k)\Phi(k, \ell+1)\mathbf{B}(\ell) \tag{5.122}$$

From (5.117), for $k \geqslant m \geqslant \ell$,

$$\prod_{i=\ell}^{k-1} \mathbf{A}(i) = \prod_{i=m}^{k-1} \mathbf{A}(i) \prod_{n=\ell}^{m-1} \mathbf{A}(n) \tag{5.123}$$

Thus, the expression for Φ becomes

$$\Phi(k,\ell) = \Phi(k,m)\Phi(m,\ell) \tag{5.124}$$

For $k = \ell$ equation (5.124) gives

$$\Phi(k,k) = \Phi(k,m)\Phi(m,\ell) = \mathbf{I}$$

For $k < \ell$ we define

$$\Phi(k,\ell) = \mathbf{A}^{-1}(k)\mathbf{A}^{-1}(k+1)\ldots\mathbf{A}^{-1}(\ell-1) \tag{5.125}$$

Provided the *inverses of* $\mathbf{A}(k)$ *exist* we conclude that

$$\Phi(m,k) = \Phi^{-1}(k.m) \tag{5.126}$$

Hence, the weighting sequence matrix (5.122) can be written as the product

$$\mathbf{H}(k, k-\ell) = \mathbf{H}(k)\mathbf{H}(\ell) \tag{5.127}$$

where

$$\mathbf{H}(k) = \mathbf{C}(k)\Phi(k,0) \tag{5.128}$$

$$\mathbf{H}(\ell) = \Phi(0,\ell+1)\mathbf{B}(\ell) \tag{5.129}$$

6
Controllability, Observability and Stability

6.1 NOTATION

In certain analytical problems involving the scalar product of two *complex* vectors the order of the vectors in the scalar product is important. To highlight this importance and track the ordered vector pairs we adopt the *Dirac bra-ket* (or $\langle\cdot|\cdot\rangle$) vector notation. However, as the need arises for clarity or brevity we will occasionally revert to the original notation.

In general for any two complex vectors $\mathbf{x}$ and $\mathbf{y}$ belonging to the same vector space their scalar product may be such that

$$\mathbf{x} \cdot \mathbf{y} \neq \mathbf{y} \cdot \mathbf{x}$$

or in Dirac notation

$$\langle \mathbf{x} | \mathbf{y} \rangle \neq \langle \mathbf{y} | \mathbf{x} \rangle$$

Since the absolute value of the scalar product is independent of the order of the factors we can write

$$\begin{aligned} \langle \mathbf{x}|\mathbf{y}\rangle &= \langle \mathbf{y}|\mathbf{x}\rangle^* \\ \langle \mathbf{x}|\lambda\mathbf{y}\rangle &= \lambda\langle \mathbf{x}|\mathbf{y}\rangle \\ \langle \lambda\mathbf{x}|\mathbf{y}\rangle &= \lambda^*\langle \mathbf{x}|\mathbf{y}\rangle \end{aligned} \tag{6.1}$$

where λ is a constant, not necessarily real, and the asterisk denotes the complex conjugate.

Consider an $n \times n$ Hermitian operator $\mathbf{A}$ having distinct eigenvalues $\lambda_1, \lambda_2, \ldots, \lambda_n$. Let the associated ket (or right) eigenvectors be $|\mathbf{e}_1\rangle, |\mathbf{e}_2\rangle, \ldots, |\mathbf{e}_n\rangle$. The corresponding eigenequations can be written as

$$\mathbf{A}|\mathbf{e}_i\rangle = \lambda_i|\mathbf{e}_i\rangle \qquad i = 1,2,\ldots,n \tag{6.2}$$

where each $|\mathbf{e}_i\rangle$ is defined by the normalized basis vectors $({}^1e_i, {}^2e_i, \ldots, {}^ne_i)$, i.e.,

$$|\mathbf{e}_i\rangle = ({}^1e_i, {}^2e_i, \ldots, {}^ne_i) \tag{6.3}$$

It can be shown that the eigenvectors $|\mathbf{e}_1\rangle, |\mathbf{e}_2\rangle, \ldots, |\mathbf{e}_n\rangle$ are linearly independent and therefore, also form a basis for the n-dimensional space. Hence any vector in the space spanned by $|\mathbf{e}_i\rangle$ ($i = 1,2,\ldots,n$) can be written in terms of the eigenvectors. For the right vector $|\mathbf{x}\rangle$ we can therefore write

$$|\mathbf{x}\rangle = \sum_{i=1}^{n} x_i\, |\mathbf{e}_i\rangle \tag{6.4}$$

The respective bra or reciprocal basis vectors corresponding to those of (6.3) are $\langle \mathbf{e}_1|, \langle \mathbf{e}_2|, \ldots. \langle \mathbf{e}_n|$. Together the two *sets* of basis vectors satisfy the orthonormality relation

$$\langle \mathbf{e}_i|\mathbf{e}_j\rangle = \delta_{ij} \tag{6.5}$$

For any two specific vectors $|\mathbf{e}_i\rangle$ and $\langle \mathbf{e}_i|$ the term

$$|\mathbf{e}_i\rangle\langle \mathbf{e}_i|$$

is called a *dyad*. Provided they have been normalized to satisfy condition (6.5) the sum of the dyads is the identity matrix, i.e.,

$$\sum_{i=1}^{n} |\mathbf{e}_i\rangle\langle\mathbf{e}_i| = \mathbf{I} \tag{6.6}$$

which is the *closure relation*.

Corresponding to the right vector (6.4) we can write for the left vector

$$\langle\mathbf{x}| = \sum_{i=1}^{n} x_i \langle\mathbf{e}_i| \tag{6.7}$$

Using identity (6.6) equations (6.4) and (6.7) can respectively be written as

$$|\mathbf{x}\rangle = \sum_{i=1}^{n} |\mathbf{e}_i\rangle\langle\mathbf{e}_i|\mathbf{x}\rangle \tag{6.8}$$

$$\langle\mathbf{x}| = \sum_{i=1}^{n} \langle\mathbf{x}|\mathbf{e}_i\rangle\langle\mathbf{e}_i| \tag{6.9}$$

The scalar product of any eigenvector $\langle\mathbf{e}_j|$ and $|\mathbf{x}\rangle$ becomes

$$\langle\mathbf{e}_j|\mathbf{x}\rangle = \sum_{i=1}^{n} \langle\mathbf{e}_j|\mathbf{e}_i\rangle x_i \tag{6.10}$$

$$= x_j$$

Essentially, the dyads of (6.6) are $n \times n$ matrices with diagonal elements ${}^{a}e_i{}^{b}e_i$. Each term in the sum is a projection. Therefore, the eigenvalue problem as formulated by (6.2) can be stated as one whereby for a given Hermitian matrix operator $\mathbf{A}$ the associated n-dimensional vector space is decomposed into a complete set of orthonormal vectors such that $\mathbf{A}$ is a linear combination of the projections (dyads). This is clearly seen by expanding (6.2), i.e., multiplying both sides from the right by $\langle\mathbf{e}_i|$ and summing:

$$\mathbf{A}|\mathbf{e}_i\rangle = \lambda_i|\mathbf{e}_i\rangle$$

$$\mathbf{A}\sum_{i=1}^{n} |\mathbf{e}_i\rangle\langle\mathbf{e}_i| = \sum_{i=1} \lambda_i|\mathbf{e}_i\rangle\langle\mathbf{e}_i|$$

$$\mathbf{A} = \sum_{i=1}^{n} \lambda_i |\mathbf{e}_i\rangle\langle\mathbf{e}_i| \tag{6.11}$$

Since any matrix operator **A** can be written as its identity

$$\mathbf{A} = \mathbf{IA} \tag{6.12}$$

it is apparent from (6.6) that matrix operators can be expanded in terms of their respective eigenvectors. Thus the following identities are possible:

$$\mathbf{A} = \mathbf{A}\sum_{i=1}^{n} |\mathbf{e}_i\rangle\langle\mathbf{e}_i| = \sum_{i=1}^{n} \lambda_i |\mathbf{e}_i\rangle\langle\mathbf{e}_i| \tag{6.13}$$

$$\mathbf{A}^{-1} = \sum_{i=1}^{n} \frac{1}{\lambda_i} |\mathbf{e}_i\rangle\langle\mathbf{e}_i| \tag{6.14}$$

$$|\mathbf{x}\rangle = \sum_{i=1}^{n} |\mathbf{e}_i\rangle\langle\mathbf{e}_i|\mathbf{x}\rangle \tag{6.15}$$

$$(s\mathbf{I} - \mathbf{A}) = \sum_{i=1}^{n} (s - \lambda_i)|\mathbf{e}_i\rangle\langle\mathbf{e}_i| \tag{6.16}$$

$$(s\mathbf{I} - \mathbf{A})^{-1} = \sum_{i=1}^{n} \frac{1}{(s - \lambda_i)} |\mathbf{e}_i\rangle\langle\mathbf{e}_i| \tag{6.17}$$

$$e^{\mathbf{A}t} = \sum_{i=1}^{n} e^{\lambda_i t} |\mathbf{e}_i\rangle\langle\mathbf{e}_i| \tag{6.18}$$

$$f(\mathbf{A}) = \sum_{i=1}^{n} f(\lambda_i)|\mathbf{e}_i\rangle\langle\mathbf{e}_i| \tag{6.19}$$

6.2 SPECTRAL DECOMPOSITION

The state equations in standard form were shown to be

$$|\dot{\mathbf{x}}\rangle = \mathbf{A}|\mathbf{x}\rangle + \mathbf{B}|\mathbf{u}\rangle$$

$$|\mathbf{y}\rangle = \mathbf{C}|\mathbf{x}\rangle + \mathbf{D}|\mathbf{u}\rangle$$

Their respective time-dependent solutions are

$$|\mathbf{x}\rangle = \Phi(t)\left[|\mathbf{x}(0)\rangle + \int_0^t e^{-\mathbf{A}\xi}\mathbf{B}|\mathbf{u}(\xi)\rangle d\xi\right] \tag{6.20}$$

$$|\mathbf{y}\rangle = \mathbf{C}\Phi(t)\left[|\mathbf{x}(0)\rangle + \int_0^t e^{-\mathbf{A}\xi}\mathbf{B}|\mathbf{u}(\xi)\rangle d\xi\right] + \mathbf{D}|\mathbf{u}(t)\rangle \tag{6.21}$$

By applying identities (6.15) and (6.18) to solutions (6.20) and (6.21) we obtain the following *spectral decompositions*:[1]

$$|\mathbf{x}\rangle = \sum_{i=1}^{n} e^{\lambda_i t}|\mathbf{e}_i\rangle\langle\mathbf{e}_i|\mathbf{x}(0)\rangle + \sum_{i=1}^{n}\int_0^t e^{\lambda_i(t-\xi)}|\mathbf{e}_i\rangle\langle\mathbf{B}^\dagger\mathbf{e}_i|\mathbf{u}(\xi)\rangle d\xi \tag{6.22}$$

$$|\mathbf{y}\rangle = \sum_{i=1}^{n} e^{\lambda_i t}\mathbf{C}|\mathbf{e}_i\rangle\langle\mathbf{e}_i|\mathbf{x}(0)\rangle + \sum_{i=1}^{n}\int_0^t e^{\lambda_i(t-\xi)}\mathbf{C}|\mathbf{e}_i\rangle\langle\mathbf{B}^\dagger\mathbf{e}_i|\mathbf{u}(\xi)\rangle d\xi + \mathbf{D}|\mathbf{u}(t)\rangle \tag{6.23}$$

where $\mathbf{B}^\dagger$ is the conjugate transpose of $\mathbf{B}$. The scalar product

$$\langle\mathbf{e}_i|\mathbf{B}\mathbf{u}(\xi)\rangle = \langle\mathbf{B}^\dagger\mathbf{e}_i|\mathbf{u}(\xi)\rangle$$

applies.

From (6.22) and (6.23) two observations are readily apparent: (1) the vectors $\langle\mathbf{B}^\dagger\mathbf{e}_i|$ may be regarded as "weights" which determine the magnitude of the effect of the input on the change in state in the $|\mathbf{e}_i\rangle$ "direction," and (2) the vector-valued, time-dependent quantities $\mathbf{e}^{\lambda_i t}|\mathbf{e}_i\rangle$ can be interpreted as the system *modes*.

6.3 CONTROLLABILITY

The system theoretic concept of controllability arises from the following two-point boundary valve problem. Given an initial state $|\mathbf{x}(0)\rangle$ at time zero and a final state $|\mathbf{x}\rangle \neq |\mathbf{x}(0)\rangle$, determine whether it is possible to find a time t and an input $|\mathbf{u}\rangle$ which take the system from its initial state at time zero to its final

1. Notationally $\mathbf{x} = |\mathbf{x}\rangle$.

state at time t, where t is finite. To formalize the definition of controllability we establish for the system *representation*

$$|\dot{\mathbf{x}}\rangle = \mathbf{A}|\mathbf{x}\rangle + \mathbf{B}|\mathbf{u}\rangle$$

$$|\mathbf{y}\rangle = \mathbf{C}|\mathbf{x}\rangle + \mathbf{D}|\mathbf{u}\rangle \tag{6.24}$$

and stipulate that at some initial time zero the system is in an allowable initial state $|\mathbf{x}(0)\rangle$. The system *representation* (6.24) is *defined*[1] *to be completely controllable if there exists a finite time* $t > 0$ *and a real input* $|\mathbf{u}\rangle$ *defined on the time interval* $(0,t)$ *such that*

$$|\mathbf{x}\rangle = e^{\mathbf{A}t}\left[|\mathbf{x}(0)\rangle + \int_0^t e^{-\mathbf{A}\xi}\mathbf{B}|\mathbf{u}(\xi)\rangle d\xi\right]$$

The necessary and sufficient conditions under which the system representation is controllable are readily deducible from (6.22), the spectral decomposition of the state. These conditions are presented below (without proof) in the form of the following theorems:[2]

Theorem 6.1. *Let the matrix* $\mathbf{A}$ *in* (6.24) *have distinct eigenvalues. Representation* (6.24) *is completely controllable if and only if for all n vectors of* (6.22) *the condition*

$$\langle \mathbf{B}^\dagger \mathbf{e}_i| \neq 0 \qquad (i = 1,2,\ldots,n) \tag{6.25}$$

is satisfied. $\langle \mathbf{B}\mathbf{e}_i| = 0$ for any i implies that the mode e_i cannot be excited. Hence the representation cannot be completely controlled.

Theorem 6.2. *Let* $\mathbf{A}$ *and* $\mathbf{B}$ *in* (6.24) *be* $n \times n$ *and* $n \times m$ *matrices, respectively. Representation* (6.24) *is completely controllable if and only if the rows of the matrix*

$$e^{-\mathbf{A}\xi}\mathbf{B}$$

are linearly independent on the time interval $(0, t)$.

1. See Zadeh and Polak, System Theory, p. 244, McGraw-Hill, New York, 1969.
2. *op. cit.*

Theorem 6.3. Let **A** *and* **B** *in* (6.24) *be* $n \times n$ *and* $n \times m$ *matrices, respectively. Representation* (6.24) *is completely controllable if and only if the* $n \times nm$ *controllability matrix* $\mathbf{Q}_c$, *where*

$$\mathbf{Q}_c = [\mathbf{B} \;\; \mathbf{AB} \;\; \mathbf{A}^2\mathbf{B} \;\; \dots \;\; \mathbf{A}^{n-1}\mathbf{B}] \tag{6.26}$$

has rank n.

As a test of the above conditions consider the simple system of Fig. 6.1. We desire to establish the controllability of the system representation. By inspection we have the simultaneous equations

$$\dot{x}_1 = x_1 + 4x_2 + u$$

$$\dot{x}_2 = 5x_2 + u$$

$$y = x_1 - x_2$$

which can be written in matrix form as

$$\begin{bmatrix} \dot{x}_1 \\ \dot{x}_2 \end{bmatrix} = \begin{bmatrix} 1 & 4 \\ 0 & 5 \end{bmatrix} \begin{bmatrix} x_1 \\ x_2 \end{bmatrix} + \begin{bmatrix} 1 \\ 1 \end{bmatrix} u$$

$$y = [1 \;\; -1] \begin{bmatrix} x_1 \\ x_2 \end{bmatrix} + u$$

Thus the system matrices are identified as

$$\mathbf{A} = \begin{bmatrix} 1 & 4 \\ 0 & 5 \end{bmatrix} \qquad \mathbf{B} = \begin{bmatrix} 1 \\ 1 \end{bmatrix} \qquad \mathbf{C} = [1 \;\; -1] \qquad \mathbf{D} = [1]$$

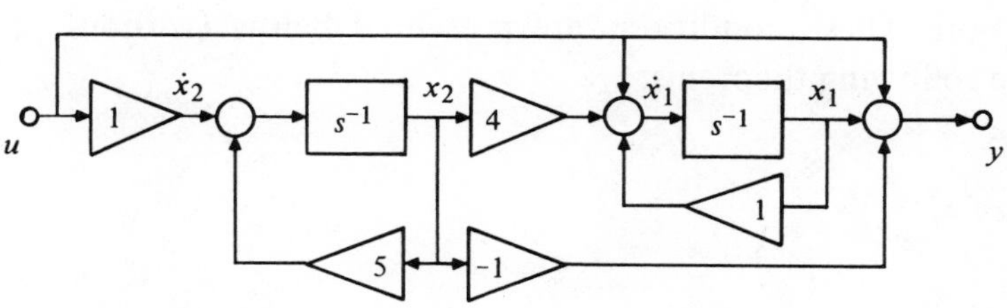

Figure 6.1

From (6.26) the controllability matrix $\mathbf{Q}_c$ is the 2×2 matrix

$$\mathbf{Q}_c = [\mathbf{B} \quad \mathbf{AB}] = \begin{bmatrix} 1 & 5 \\ 1 & 5 \end{bmatrix}$$

Clearly, the columns $\mathbf{B}$ and $\mathbf{AB}$ of $\mathbf{Q}_c$ are linearly dependent. Therefore $\mathbf{Q}_c$ does not have rank n (=2). Hence, the system representation for the above example is not completely controllable. However, with a slight modification to $\mathbf{A}$ (i.e., changing the amplification of state component x_2), such that

$$\mathbf{A} = \begin{bmatrix} 1 & 5 \\ 0 & 5 \end{bmatrix}$$

the columns of $\mathbf{Q}_c$ become linearly independent. As a result the system representation for the above example is now completely controllable. We have

$$\mathbf{Q}_c = \begin{bmatrix} 1 & 6 \\ 1 & 5 \end{bmatrix}$$

6.4 OBSERVABILITY

Changes in system state as discussed above are not directly observable. Usually what is observed is the input $|\mathbf{u}\rangle$ and its detectable effect on the system output $|\mathbf{y}\rangle$. This physical realization leads to the following definition of observability. We establish the system representation as (6.24). Let the system be in an allowable initial state $|\mathbf{x}(0)\rangle$ at time zero. And let the input at time zero be $|\mathbf{u}\rangle \equiv 0$. The system representation (6.24) is *defined*[1] *to be completely observable if for every initial state at time zero there exists a finite time t such that knowledge of the output* $|\mathbf{y}\rangle$ *over the time interval* $(0,t)$ *is sufficient to determine the initial state* $|\mathbf{x}(0)\rangle$.

A set of necessary and sufficient conditions under which the system representation (6.24) is observable is seen from (6.23), the output spectral decomposition. These conditions are presented below (without proof) in the form of the following theorems:[2]

1. *op. cit.*
2. *op. cit.*

Theorem 6.4. *Let the matrix* **A** *in* (6.24) *have distinct eigenvalues. Representation* (6.24) *is completely observable if and only if for the n vectors of* (6.23) *the condition*

$$\mathbf{C}|\mathbf{e}_i\rangle \neq 0 \tag{6.27}$$

is satisfied. Obviously, if $\mathbf{C}|\mathbf{e}_i\rangle = 0$ for some i, then initial states of the form $|\mathbf{x}(0)\rangle = a|\mathbf{e}_i\rangle$ would give zero outputs for zero input and these states would be indistinguishable from the zero-state $|\mathbf{x}\rangle = 0$. (Note that in concert with (6.25) the system representation is completely controllable if and only if every mode of its dual representation is observable.)

Theorem 6.5. *Let the matrices* **A**, **B**, **C** *and* **D** *in* (6.24) *be, respectively,* $n \times n$, $n \times m$, $k \times n$ *and* $k \times m$. *Representation* (6.24) *is completely observable if and only if the observability matrix* $\mathbf{Q}_0$, *where*

$$\mathbf{Q}_0^\dagger = [\mathbf{C}^\dagger \, \mathbf{A}^\dagger \mathbf{C}^\dagger \ldots (\mathbf{A}^{n-1})^\dagger \mathbf{C}^\dagger] \tag{6.28}$$

has rank n. (The dagger denotes the conjugate transpose.)

As a test of the above criteria we return to the simple system of Fig. 6.1. The matrix **C** is

$$\mathbf{C} = [1 \quad -1]$$

By (6.28) the observability matrix $\mathbf{Q}_0$ is

$$\mathbf{Q}_0 = \begin{bmatrix} 1 & -1 \\ 1 & -1 \end{bmatrix}$$

Clearly the columns of $\mathbf{Q}_0$ are linearly dependent and the rank of $\mathbf{Q}_0$ is less than n (=2). Hence the representation for the system of Fig. 6.1 is not completely observable. However, with a slight modification of **C** (i.e., changing the amplification of the output x_2) such that

$$\mathbf{C} = [1 \quad 1]$$

the columns of $\mathbf{Q}_0$ are made linearly independent and the system representation is made completely observable.

6.5 STABILITY-UNFORCED SYSTEM

The system theoretic concepts of controllability and observability are direct results of the spectral decomposition of the state and output vectors. Both concepts appeal to our intuition. A third concept, somewhat related to the other two and also fundamental to the qualitative analysis of dynamical systems, is the concept of system *stability*. In defining stability we refer to the system state and its behavior with time. By definition an equilibrium state is one whereby

$$|\dot{\mathbf{x}}\rangle = 0$$

$$|\mathbf{x}\rangle = \overline{\theta}$$

From Newton's laws of motion an unforced system that is initially in an equilibrium state will remain in that state indefinitely unless acted upon by an external force, after which time one of several mutually exclusive things can happen. Consider the external force to be an impulse. Then, (1) the system can be displaced from equilibrium and, by internal properties, returned to equilibrium within a small time interval. In this case the state is said to be *stable*. (2) The state can be displaced a finite "distance" from equilibrium and remain at the displaced position for all time. In this case the state is said to be *unstable* but *bounded*. Or (3) the state vector can grow indefinitely with time in which case the state is both *unstable* and *unbounded*.

The bounded aspects of system or state stability can be refined mathematically by introducing the idea of a neighborhood surrounding an equilibrium state. What is intended by a neighborhood in this instance is a finite, fixed, arbitrarily small displacement ϵ in the t_0 time plane which surrounds the end-point of the equilibrium state vector $\mathbf{x}_\epsilon$ (Fig. 6.2). We assume that the system output is the state and that the equilibrium state in question is the zero-state $|\mathbf{x}\rangle = \overline{\theta} = 0$. Accordingly, we *define* the concept of *stability of the zero-state* or *zero-input stability*. Roughly, the zero-state $|\mathbf{x}\rangle = 0$ is stable if for every initial

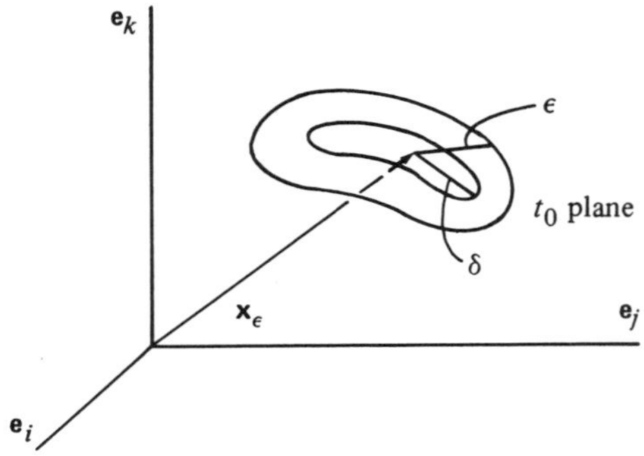

Figure 6.2 Neighborhood surrounding an equilibrium state.

state $|\mathbf{x}(0)\rangle$ sufficiently close to zero the corresponding *free motion* $\mathbf{x}(t_0,\mathbf{x}(0),t)$ remains close to zero for all $t \geqslant t_0$ (Fig. 6.3). More precisely, *we define the zero-state* $|\mathbf{x}\rangle = 0$ *to be stable in the sense of Lyapunov if for any* t_0 *and any* $\epsilon > 0$ *there exists a* $\delta > 0$, *dependent on* ϵ *and* t_0, *such that*

$$|\mathbf{x}(0)| < \delta \Rightarrow |\mathbf{x}| < \epsilon \tag{6.29}$$

Condition (6.29) implies that the free motion trajectory $\mathbf{x}(t_0,\mathbf{x}(0),t)$ remains in the cylinder of radius ϵ for all time.

Additionally, we require that the system response eventually go to zero. Our above definition of stability in the sense of Lyapunov does not cover this requirement. Therefore, we *define* the stability of the zero-state $|\mathbf{x}\rangle = 0$ to be *asymptotically stable* if (a) *it is stable in the sense of Lyapunov*, and (b) *for any* t_0 $|\mathbf{x}(0)\rangle$ *is sufficiently close to* 0 *such that the free motion trajectory* $\mathbf{x}(t_0,\mathbf{x}(0),t) \to 0$ *as* $t \to \infty$.

We have seen previously that for the *unforced* system the solution to the dynamic state equation is

$$|\mathbf{x}\rangle = \Phi(t)|\mathbf{x}(0)\rangle$$

$$\Phi(t) = e^{\mathbf{A}t} \tag{6.30}$$

Thus it is possible to represent the state at a time subsequent to t_0 by a linear transformation (involving the fundamental matrix Φ) of the initial state. Therefore, for such systems it should be possible to determine the conditions for stability which depend only on Φ. Accordingly, in support of the *definitions* of *zero-state stability* the following *theorems*[1] apply:

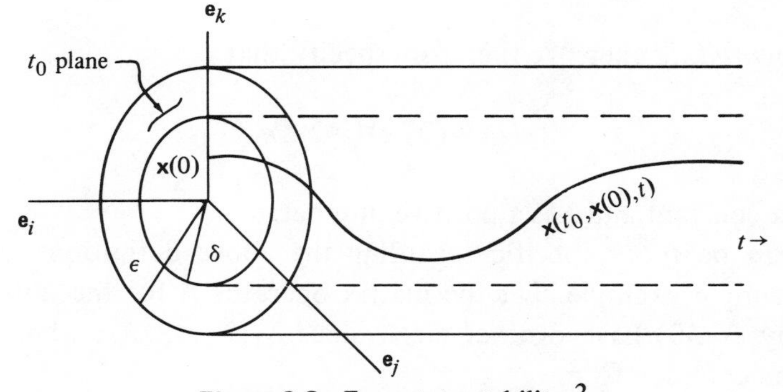

Figure 6.3 Zero-state stability.[2]

1. The above theorems are presented without proof. For a demonstration of their respective proofs see, for example, Zadeh and Desoer, Linear Systems Theory, pp. 379-391, pp. 498-503, McGraw-Hill, New York, 1963.
2. Notationally $|\mathbf{x}\rangle = \mathbf{x}$ and the two notations will be used interchangeably where brevity is required.

Theorem 6.6. *The system is stable in the sense of Lyapunov implies (and is implied by) there exists a finite constant K which may depend on t_0 such that*

$$|\Phi(t,t_0)| \leqslant K \qquad \text{(for all } t \geqslant t_0) \tag{6.31}$$

Theorem 6.7. *The system is asymptotically stable if*

(a) $$|\Phi(t,t_0)| \leqslant K \qquad \text{(for all } t \geqslant t_0)$$

and (b) $$\lim_{t\to\infty} |\Phi(t,t_0)| = 0 \qquad \text{(for all } t_0) \tag{6.32}$$

Theorem 6.8. *The system is stable in the sense of Lyapunov implies (and is implied by)*

(a) *All the eigenvalues of the constant matrix* **A** *have non-real parts, and*

(b) *those eigenvalues of* **A** *that lie on the imaginary axis are simple zeros of the minimum polynomial of* **A**.

Theorem 6.9. *The system is asymptotically stable if and only if all the eigenvalues of* **A** *have negative real parts.*

By theorems 6.7 and 6.9 above we assert that for fixed systems described by (6.30) asymptotic stability implies that

$$|\mathbf{x}(t_0, \mathbf{x}(0), t)| \to 0$$

in an *exponential* manner. We therefore specify that

$$|\mathbf{x}(t_0, \mathbf{x}(0), t)| \leqslant Ce^{-\lambda t} \tag{6.33}$$

where C is a constant and λ is a positive number.

We can be more specific regarding the above definitions and theorems through a simple example. Let the matrix operator **A** for the unforced system governed by (6.30) have distinct eigenvalues $\lambda_1, \lambda_2, \ldots, \lambda_n$, where each eigenvalue has the complex form

$$\lambda_1 = \sigma_1 + j\omega_1$$
$$\lambda_2 = \sigma_2 + j\omega_2$$
$$\cdots\cdots\cdots\cdots\cdots$$
$$\lambda_n = \sigma_n + j\omega_n$$

The real parts of each eigenvalue can be ordered as $\sigma_1 > \sigma_2 > \ldots > \sigma_n$. From equation (5.46) the state transition matrix $\mathbf{\Phi}$ can be written as

$$\mathbf{\Phi} = \sum_{i=1}^{n} e^{\lambda_i t} \sum_{r=1}^{m_p} \mathbf{A}_{ri} \frac{t^{r-1}}{(r-1)!}$$

where the $\mathbf{A}_{ri}$ are matrix coefficients of the polynomial expansion and m_p is the root multiplicity or degeneracy of the eigenvalues. For the unforced system we consider three situations: (1) $\sigma_i < 0$, (2) $\sigma_i > 0$, and (3) $\sigma_i = 0$. In each case we refer to equation (5.46).

Case (1): $\sigma_i < 0$.

We see that for $t \geqslant t_0$ the absolute value of $\mathbf{\Phi}$ remains finite. The condition of Theorem 6.6 is satisfied. Further,

$$\lim_{t \to \infty} \mathbf{\Phi}(t_0, t) = 0$$

thereby also satisfying the conditions of Theorem 6.7. It necessarily follows that Theorem 6.8 and 6.9 are also satisfied. Clearly then for the case where the eigenvalues have negative real parts the state vector $|\mathbf{x}\rangle \to 0$ as $t \to \infty$. The system representation is *asymptotically stable*.

Case (2): $\sigma_i > 0$.

The time limiting value of the transition matrix $\mathbf{\Phi}$ in this case is

$$\lim_{t \to \infty} \mathbf{\Phi}(t_0, t) = \infty$$

Therefore the state vector $|\mathbf{x}\rangle \to \infty$ as $t \to \infty$. Hence, for the case where the system eigenvalues have positive real parts the system representation is *unstable*.

Case (3): $\sigma_i = 0$.

When the system eigenvalues are purely imaginary two situations arise:

(a) The system has *simple roots*, in which case the $\mathbf{A}_{ri}$ are constant matrices and $\mathbf{\Phi}$ remains finite. Thus in accordance with Theorem 6.6 the system representation is *stable*.

(b) The system has *multiple roots*, in which case the $\mathbf{A}_{ri}$ are polynomials in t with matrix coefficients. Clearly

$$\lim_{t \to \infty} \mathbf{\Phi}(t_0, t) = \infty$$

Hence, the system representation is *unstable*.

6.6 STABILITY-FORCED SYSTEM

Recalling the state dynamical equation and its solution we have for the forced system

$$|\dot{\mathbf{x}}\rangle = \mathbf{A}|\mathbf{x}\rangle + \mathbf{B}|\mathbf{u}\rangle$$

$$|\mathbf{x}\rangle = \Phi(t)\left[|\mathbf{x}(0)\rangle + \int_0^t \Phi(\xi)\mathbf{B}|\mathbf{u}(\xi)\rangle d\xi\right]$$

For discussion purposes we specify that the system is initially at rest. Let the initial state at time zero be the zero-state, i.e., $|\mathbf{x}(0)\rangle = 0$, after which time the system is perturbed by an impulse. We examine the system response as it relates to the concept of stability. For this situation the solution to the dynamical equation becomes

$$|\mathbf{x}\rangle = \int_0^t \Phi(t-\xi)\mathbf{B}|\mathbf{u}(\xi)\rangle d\xi$$

$$= \int_0^t \mathbf{G}|\mathbf{u}(\xi)\rangle d\xi \tag{6.34}$$

where

$$\mathbf{G} = \Phi(t-\xi)\mathbf{B} \tag{6.35}$$

The elements of the matrix **G** are of the form

$$g_{ij} = \sum_{k=1}^{n} \varphi_{ik} b_{kj} \tag{6.36}$$

Since the vector $|\mathbf{u}\rangle$ is a column matrix the relationship between the elements of the integrand and the state vector is established as

$$x_i = \sum_j g_{ij} u_j$$

$$= \sum_j \sum_k \varphi_{ik} b_{kj} u_j \qquad (i = 1, 2, \ldots, n) \tag{6.37}$$

Expressing the elemental input as an impulse

$$u_j = \delta(t - \xi)$$
$$u_i = 0 \qquad \text{for } i \neq j$$

the components of the state vector become

$$x_i = g_{ij} \qquad (i = 1, 2, \ldots, n)$$

Thus **G** can be interpreted as the matrix of impulse responses of the state. For such a system we *define* stability as follows: *A forced system is stable relative to a set of bounded inputs if and only if the state is bounded for all inputs* $|\mathbf{u}\rangle(t)$ *in the set for all* $t \geq t_0$. Accordingly we require that

$$|\mathbf{x}(t)| \leq K < \infty \tag{6.38}$$

where K is a finite constant. In support of the above definition the stability criterion for a forced system is established by the following theorem:[1]

Theorem 6.10. *A forced system is stable with respect to a set of bounded inputs if and only if*

$$\int_{t_0}^{t} |g_{ij}| d\xi \leq K_{ij} < \infty \tag{6.39}$$

for every t_0 and all $t > t_0$.

We illustrate through a simple example the concept of stability as it applies to a forced system. Consider the representation of the system in Fig. 6.2. The system matrices are

$$\mathbf{A} = \begin{bmatrix} 1 & 4 \\ 0 & 5 \end{bmatrix} \qquad \mathbf{B} = \begin{bmatrix} 1 \\ 1 \end{bmatrix}$$

$$\mathbf{C} = [1 \quad -1] \qquad \mathbf{D} = [1]$$

1. *op. cit.*

The state transition matrix is

$$\Phi = e^{\mathbf{A}t} = \mathbf{I} + \mathbf{A}t + \frac{(\mathbf{A}t)^2}{2!} + \ldots$$

$$= \begin{bmatrix} 1 & 0 \\ 0 & 1 \end{bmatrix} + \begin{bmatrix} 1 & 4 \\ 0 & 5 \end{bmatrix} t + \begin{bmatrix} 1 & 24 \\ 0 & 25 \end{bmatrix} \frac{t^2}{2!} + \ldots$$

We desire to examine in closed form the time dependency of the fundamental matrix. This can readily be accomplished for simple systems by converting to the frequency domain then back to the time domain. Thus

$$\mathcal{L}\{\Phi(t)\} = [s\mathbf{I} - \mathbf{A}]^{-1}$$

$$= \begin{bmatrix} s-1 & -4 \\ 0 & s-5 \end{bmatrix}^{-1}$$

$$= \begin{bmatrix} \dfrac{1}{s-1} & \dfrac{4}{(s-1)(s-5)} \\ 0 & \dfrac{1}{(s-5)} \end{bmatrix}$$

Converting back to the time domain the fundamental matrix becomes

$$\Phi(t) = \begin{bmatrix} e^t & e^{5t} - e^t \\ 0 & e^{5t} \end{bmatrix}$$

Hence the product $\mathbf{G} = \Phi\mathbf{B}$ is

$$\mathbf{G} = \begin{bmatrix} e^{5t} \\ e^{5t} \end{bmatrix}$$

It is readily apparent that for the system of Fig. 6.1 the positive nature of the elements of Φ and $\mathbf{B}$ will cause all the non-zero terms to grow beyond bounds as t becomes infinite. Therefore, the system representation for the example chosen, under both forced and unforced conditions, is unstable. This conclusion should be no surprise. Earlier it was shown that the representation for the system of Fig. 6.1 was not completely controllable and not completely observable. However, with slight modifications we were able to render the representation as completely controllable and completely observable. We now logically inquire as to whether or not similar modifications can be made in the case of system stability—under forced and unforced conditions.

In the case of the unforced system the conditions for stability are manifest in equations (6.31) through (6.33). Equations (6.38) and (6.39) represent the conditions for stability of the forced system. All of these conditions can be satisfied in a variety of ways; the objective being to maintain a finite state vector for bounded inputs for all $t \geqslant t_0$. An unstable representation of an unforced system does not necessarily mean an unstable representation of that same system when an input is applied. The terms giving rise to the instability may not be excited by the input. Nor does it necessarily follow that when the conditions for stability are met then the system is completely observable and/or completely controllable. Similarly, a stable representation of an unforced system does not necessarily imply a stable representation of the forced system. We demonstrate these important facts through a simple example. Referring to the system of Fig. 6.1, we render the system stable by changing the representation to that shown in Fig. 6.4. Essentially, the changes are in the magnitudes of signal amplification. The modified system matrices become[1]

$$\mathbf{A} = \begin{bmatrix} 1 & 1 \\ 0 & -1 \end{bmatrix} \qquad \mathbf{B} = \begin{bmatrix} 1 \\ -2 \end{bmatrix}$$

$$\mathbf{C} = [1 \quad 1] \qquad \mathbf{D} = [1]$$

The corresponding state transition matrix $\mathbf{\Phi}$ and impulse response matrix $\mathbf{G}$ are, respectively,

$$\mathbf{\Phi} = \begin{bmatrix} e^t & \frac{1}{2}(e^t - e^{-t}) \\ 0 & e^{-t} \end{bmatrix} \qquad \mathbf{G} = \begin{bmatrix} e^{-t} \\ -2e^{-t} \end{bmatrix}$$

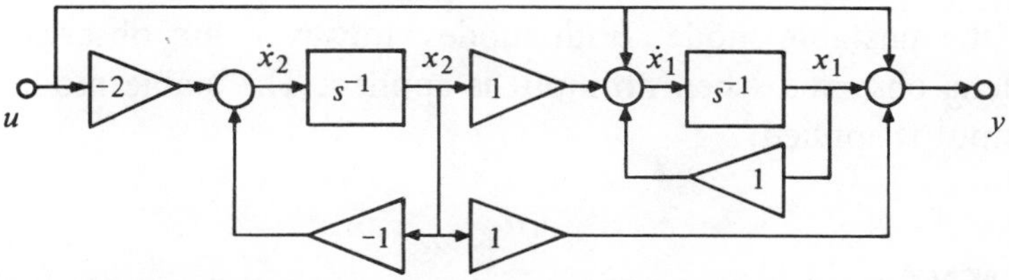

Figure 6.4 System wherein the input does not excite the unstable mode.

1. After Schwarz and Friedland, Linear Systems, p. 381, McGraw-Hill, New York, 1965.

Due to the positive nature of the elements φ_{11} and φ_{12} we have

$$\lim_{t\to\infty} |\Phi| = \infty$$

Thus, the representation for the system of Fig. 6.4 is unstable when no input is applied. On the other hand the impulse response contains the diminishing terms e^{-t} in both elements g_1 and g_2. Therefore the integrals

$$\int_{t_0}^{\infty} |g_1| d\xi \qquad \text{and} \qquad \int_{t_0}^{\infty} |g_2| d\xi$$

both remain finite for all finite $t \geqslant t_0$. Consequently, the representation of the system of Fig. 6.4 is stable when an input is applied. This apparent paradox is resolved when one examines the controllability and observability matrices for the system, under both forced and unforced conditions:

Unforced system	Forced system
$\mathbf{u} = 0$	
$\mathbf{B} = 0$	$\mathbf{Q}_c = \begin{bmatrix} 1 & -1 \\ -2 & 2 \end{bmatrix}$
$\mathbf{Q}_c = 0$	
$\mathbf{Q}_0 = \begin{bmatrix} 1 & 1 \\ 1 & 0 \end{bmatrix}$	$\mathbf{Q}_0 = \begin{bmatrix} 1 & 1 \\ 1 & 0 \end{bmatrix}$

Conclusions:

Unforced system	Forced system
1. unstable	1. stable
2. not completely controllable	2. not completely controllable
3. completely observable	3. completely observable

Thus, to resolve the above paradox it is concluded that the system input does not excite the unstable mode. Both modes, however, are observable. The unstable mode is observed when no input is applied. The stable mode is observed when the input is applied.

6.7 COMMENT

The central idea of the past discussion has been system representation in state-space. The algebra of rational functions (polynomials), linear vector spaces, and the methods of Fourier and Laplace transforms were the principal analytical tools. It was assumed that knowledge of the system matrix operators and variables was complete. However, it must be recognized that for *large* systems this assumption is short-lived. For large systems complete identification of all the variables and their operator elements is *not* practical, and therein lies the problem of applying the linear theory to practical problems dealing with systems of significant size. Consequently, a deterministic analysis for such systems gives way to the more realizable probabilistic or statistical analysis. Accordingly, the reader is referred to the methods of Markov chains, diakoptics, sparse matrices, fuzzy sets, statistical mechanics, etc., which bear on this problem.

7
Statistical Systems-Signals in Noise

7.1 INTRODUCTION

It is important to recognize the practical limitations of the theory developed this far. Certainly, a deterministic analysis is possible for simple differential systems, where all the descriptors satisfying the input-output-state relations are known. However, for large complex systems all the state variables, the elements of the fundamental matrix, etc., are not known nor can they even be defined in some cases. The theory addresses a small part of the real (nonlinear) world. Consequently, a deterministic analysis of many real systems is either not a simple task or not feasible using the linear theory developed thus far.

As a method for augmenting the theory we turn to statistical or probabilistic analysis. The input-output relations are treated as random processes. We will look at the statistical properties of each process, where the relations between these properties will form the basis of the analysis. The most important of these are the *mean value* and *correlation function*. Our attention will specifically focus on the filtering problem and stochastic processes, i.e., on signals $\underline{u}(t)$ that are described by their averages rather than signals $u(t)$ that are described by their point properties. Appendix F contains a short summary of some of the probabilistic expressions which will be used in the sequel.

7.2 AVERAGES (MEAN VALUE) AND CORRELATION FUNCTION

As a prelude to the discussion to follow we consider two different ways to represent an average. They are the *time average* and *ensemble average*. Both lead to the same results. The time average $\langle u(t)\rangle$ of the signal $u(t)$ is defined as

$$\langle u(t)\rangle = \lim_{T\to\infty} \frac{1}{2T}\int_{-T}^{T} u(t)\,dt \tag{7.1}$$

where T is the time interval $(-t,t)$. This limit is a number associated with the functional $u(t)$. On the other hand the average of a random signal is interpreted differently. Assume that the set $\{\underline{u}(t)\}$ is a stochastic process as defined in Appendix F. For a given time t, the sample function $\underline{u}(t)$ is a random process resulting in the random variable ζ. The expected value of the random variable will be denoted as $E\{\underline{u}(t)\}$; it is the *ensemble average* of the random process $\{\underline{u}(t)\}$. We have

$$E\{\underline{u}(t)\} = E\{\underline{u}(\zeta,t)\} \tag{7.2}$$

$$= \overline{\underline{u}(t)} \tag{7.3}$$

$$= \int_{-\infty}^{\infty} af(a;t)\,dt \tag{7.4}$$

where $f(a;t)$ is the probability density function associated with the value a of the random variable, and $f(a;t)da$ is the probability that a will be found in $a \pm da$. In conjunction with the above it follows that

$$\int_{-\infty}^{\infty} f(a;t)da = 1 \tag{7.5}$$

In general $\langle u(t)\rangle$ and $E\{\underline{u}(t)\}$ are unrelated. $\langle u(t)\rangle$ is a constant, whereas $E\{\underline{u}(t)\}$ depends on t. However, if the process is *stationary* (see Appendix F) and satisfies certain ergodic conditions, then the time average $\langle u(t)\rangle$ performed on each member of the set is almost certainly a constant equal to $E\{\underline{u}(t)\}$, i.e.,

$$\langle u(t)\rangle = E\{\underline{u}(t)\} \tag{7.6}$$

The correlation function $R(\xi)$ for the process $\underline{u}(t)$ is defined by (F.8) as

$$R(\xi) = \overline{\underline{u}(t-\xi)\underline{u}(t)} \tag{7.7}$$

It is assumed that $\underline{u}(t)$ is a stationary stochastic process where $\overline{\underline{u}(t)} = 0$. Expression (7.7) characterizes the statistical relationship between the values of $\underline{u}(t)$ at times t and $t - \xi$, where ξ can be positive or negative. It is tacitly implied that as ξ increases the statistical coupling between $\underline{u}(t)$ and $\underline{u}(t-\xi)$ becomes weaker.

Since $\underline{u}(t)$ is stationary the mean values $\overline{\underline{u}(t)}$, $\overline{\underline{u}^2(t)}$, $\overline{\underline{u}(t-\xi)\underline{u}(t)}$, etc., are independent of t and depend only on ξ. Therefore,

$$\overline{\underline{u}(t-\xi)\underline{u}(t)} = \overline{\underline{u}(-\xi)\underline{u}(0)} = \overline{\underline{u}(0)\underline{u}(\xi)} \tag{7.8}$$

or

$$R(\xi) = R(-\xi) \tag{7.9}$$

For two processes, say $\underline{u}(t)$ and $\underline{y}(t)$, the *cross correlation* function $R_{uy}(\xi)$ is defined as

$$R_{uy}(\xi) = \underline{u}(t)\underline{y}(t-\xi) \tag{7.10}$$

Expression (7.10) is unlike (7.7) in the sense that it is not an even function:

$$R_{uy}(\xi) = R_{yu}(-\xi) \tag{7.11}$$

where

$$R_{yu}(-\xi) = \underline{y}(t)\underline{u}(t-\xi) = \underline{y}(t-\xi)\underline{u}(t) \tag{7.12}$$

In a manner similar to defining the average or mean value two different ways we examine the time average alternative to (7.7). The process $\underline{u}(t)$ is said to be *ergodic* if all its statistics can be determined from a single sample function of $\underline{u}(t)$. Consequently, for ergodic processes the (time average) correlation function $R(\xi)$ is defined as, (see (F.17)),

$$R(\xi) = \lim_{T\to\infty} \frac{1}{2T} \int_{-t}^{t} \underline{u}(t-\xi)\underline{u}(t)\,d\xi \tag{7.13}$$

$$R(\xi) = E\{\underline{u}(t-\xi)\underline{u}(t)\} \qquad (7.14)$$

$$= \overline{\underline{u}(t-\xi)\underline{u}(t)} \qquad (7.15)$$

7.3 FILTERING PROBLEM

The problem of filtering can be stated as follows: Let the time dependent input, applied at the system (filter) input terminal, be represented as the random process $\underline{u}(t)$ (see Figure 7.1). It is the sum of a useful signal $\underline{s}(t)$ and noise $\underline{n}(t)$, both of which are random processes:

$$\underline{u}(t) = \underline{s}(t) + \underline{n}(t) \qquad (7.16)$$

The system H operates on the input resulting in the output $\underline{y}(t)$. We have

$$\underline{y}(t) = H[\underline{u}(t)]$$

$$= H[\underline{s}(t) + \underline{n}(t)] \qquad (7.17)$$

The problem is to choose H such that it reproduces some meaningful function, say $\underline{m}(t)$, with least possible error. The function $\underline{m}(t)$ is the transformed version of the signal $\underline{s}(t)$:

$$\underline{m}(t) = H[\underline{s}(t)] \qquad (7.18)$$

In simple filtering H is the unit operator 1, in which case (7.18) reduces to

$$\underline{n}(t) = \underline{s}(t) \qquad (7.19)$$

The instantaneous error of reproduction $\underline{\epsilon}(t)$ is the difference

$$\underline{\epsilon}(t) = \underline{y}(t) - \underline{m}(t) \qquad (7.20)$$

$$= H[\underline{s}(t) + \underline{n}(t)] - \underline{m}(t) \qquad (7.21)$$

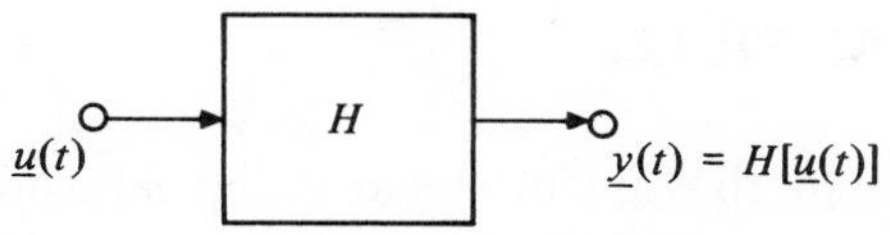

Figure 7.1

Clearly, $\underline{\epsilon}(t)$ is also a random process. The intensity of the fluctuations in the process is characterized by their *mean square value*

$$E\{\underline{\epsilon}^2(t)\} = \overline{\underline{\epsilon}^2(t)} \tag{7.22}$$

In the absence of noise, where $\underline{n}(t) = 0$, our interest will be in predicting the value of $s(t)$ after a time interval ξ:

$$\underline{m}(t) = H[s(t+\xi)] \tag{7.23}$$

The prediction is made by considering past behavior of $\underline{s}(t)$ and exploiting the statistical properties of the random process.

The discussion to follow will focus exclusively on *linear filters*. For such filters the input-output relations can be written as

$$\underline{y}(t) = H[\underline{u}(t)]$$

$$= \int_{-\infty}^{\infty} h(\xi)\underline{u}(t-\xi)d\xi \tag{7.24}$$

$$= \int_{-\infty}^{\infty} h(t-\xi)\underline{u}(\xi)d\xi \tag{7.25}$$

where $h(t)$ is the weighting function or impulse response of H. Filters characterized by either (7.24) or (7.25) are time invariant or stationary filters. The analysis will be noise suppression by linear filtering. Specifically, the filter will be the type where $\underline{u}(t)$ is stored for a time ξ then processed. It appears at the output as $\underline{y}(t)$. (Such systems resemble computers which store input data, process it, then deliver the processed data to an output terminal.) To form $\underline{y}(t)$ the values $\underline{u}(\xi)$ for all ξ are used. We seek the best system, characterized by H, where $\overline{\epsilon^2}$ is a minimum.

As a matter of convenience we will use u to mean $\underline{u}$ and similarly for all other quantities of the process. Where clarification is required the context will so state.

7.4 OPTIMUM LINEAR FILTER

The accuracy with which the useful signal is reproduced is determined by the *mean square error* $\overline{\epsilon^2}$, where ϵ is defined by (7.20). The filter H which minimizes the mean square error, wherein the output reproduces the meaningful $m(t)$, is called an *optimum* filter. To determine H we first establish from (7.24) the input-output relationship for the optimum filter:

$$y(t) = \int_{-\infty}^{\infty} h(\xi)u(t-\xi)d\xi$$

The corresponding error function is, from (7.20),

$$\epsilon(t) = \int_{-\infty}^{\infty} h(\xi)u(t-\xi)d\xi - m(t) \tag{7.26}$$

Squaring (7.26) gives

$$\begin{aligned} \epsilon^2(t) &= \left[\int_{-\infty}^{\infty} h(\xi)u(t-\xi)d\xi\right]^2 - 2m(t)\int_{-\infty}^{\infty} h(\xi)u(t-\xi)d\xi + m^2(t) \\ &= \iint_{-\infty}^{\infty} h(\xi)h(\xi')u(t-\xi)u(t-\xi')d\xi d\xi' \\ &\quad - 2\int_{-\infty}^{\infty} h(\xi)m(t)u(t-\xi)d\xi + m^2(t) \end{aligned} \tag{7.27}$$

where the square of the first term on the right-hand side was transformed into a double integral. Forming the mean of (7.27) we have

$$\begin{aligned} \overline{\epsilon^2(t)} &= \iint_{-\infty}^{\infty} h(\xi)h(\xi')\overline{u(t-\xi)u(t-\xi')}d\xi d\xi' \\ &\quad - 2\int_{-\infty}^{\infty} h(\xi)\overline{m(t)u(t-\xi)}d\xi + \overline{m^2(t)} \end{aligned} \tag{7.28}$$

From definitions (7.7) and (7.10) we introduce the autocorrelation function $R_u(\xi)$ and the cross-correlation function $R_{mu}(\xi)$, where

$$R_u(\xi) = \overline{u(t)u(t-\xi)} \tag{7.29}$$

$$R_u(\xi - \xi') = \overline{u(t-\xi)u(t-\xi')} \tag{7.30}$$

$$R_{mu}(\xi) = \overline{m(t)u(t-\xi)} \tag{7.31}$$

We note that the autocorrelation function

$$R_m(\xi) = \overline{m(t)m(t-\xi)} \tag{7.32}$$

for $\xi = 0$ reduces to

$$R_m(0) = \overline{m^2(t)} \tag{7.33}$$

Thus, the mean square error for the optimum filter can be written as

$$\overline{\epsilon^2} = \iint_{-\infty}^{\infty} h(\xi)h(\xi')R_u(\xi - \xi')d\xi d\xi' - 2\int_{-\infty}^{\infty} h(\xi)R_{mu}(\xi)d\xi + R_m(0). \tag{7.34}$$

Equation (7.34) clearly shows the dependency of the mean square error on the impulse response functions and the correlation functions. It can be shown that $\overline{\epsilon^2}$ has its minimum value if and only if the impulse response $h(t)$ is a solution to the equation

$$\int_{-\infty}^{\infty} h(\xi')R_u(\xi - \xi')d\xi' = R_{mu}(\xi) \tag{7.35}$$

Thus, it now remains to solve (7.35) for $h(\xi)$.

The solution to (7.35) will be more convenient with the following definitions. Given the correlation function $R(\xi)$ we introduce its Fourier transform $S(\omega)$, where

$$S(\omega) = \mathcal{F}\{R(\xi)\}$$

$$= \int_{-\infty}^{\infty} e^{-j\omega\xi}R(\xi)d\xi \tag{7.36}$$

Inverting (7.36) gives

$$R(\xi) = \frac{1}{2\pi}\int_{-\infty}^{\infty} e^{j\omega\xi}S(\omega)d\omega \tag{7.37}$$

The function $S(\omega)$ is the *power spectral density*, the meaning of which will become clearer later. Similarly, we introduce the transform of the impulse response $h(t)$:

$$H(\omega) = \mathcal{F}\{h(\xi)\}$$

$$= \int_{-\infty}^{\infty} e^{-j\omega\xi} h(\xi) d\xi \tag{7.38}$$

Its inverse is

$$h(\xi) = \frac{1}{2\pi}\int_{-\infty}^{\infty} e^{j\omega\xi} H(\omega) d\omega \tag{7.39}$$

The function $H(\omega)$ is called the system *transfer function*. We can now begin to solve (7.35) for $h(\omega)$.

Multiplying both sides of (7.32) by $e^{-j\omega\xi}$ and integrating with respect to ξ gives

$$\int_{-\infty}^{\infty} e^{-j\omega\xi} d\xi \int_{-\infty}^{\infty} h(\xi') R_u(\xi - \xi') d\xi' = \int_{-\infty}^{\infty} e^{-j\omega\xi} R_{mu}(\xi) d\xi \tag{7.40}$$

Letting the variable $\xi - \xi' = t$ the left-hand side of (7.40) becomes

$$\int_{-\infty}^{\infty} e^{-j\omega\xi} d\xi \int_{-\infty}^{\infty} h(\xi') R_u(\xi - \xi') d\xi' = \int_{-\infty}^{\infty} e^{-j\omega\xi'} h(\xi') d\xi' \int_{-\infty}^{\infty} e^{-j\omega t} R_u(t) dt$$

$$= H(\omega) S_u(\omega)$$

whereas the right-hand side is equal to $S_{mu}(\omega)$. Equation (7.40) therefore becomes

$$H(\omega) S_u(\omega) = S_{mu}(\omega)$$

Thus, the transfer function of the optimum filter is simply

$$H(\omega) = \frac{S_{mu}(\omega)}{S_u(\omega)} \tag{7.41}$$

i.e., it is the ratio of the spectral densities $S_{mu}(\omega)$ and $S_u(\omega)$.

Substituting equation (7.35) into (7.34) the mean square error for the optimum filter becomes

$$\overline{\epsilon^2} = R_m(0) - \iint_{-\infty}^{\infty} h(\xi)h(\xi')R_u(\xi - \xi')d\xi d\xi' \qquad (7.42)$$

In view of transforms (7.36) through (7.39) we can write

$$R_m(0) = \frac{1}{2\pi}\int_{-\infty}^{\infty} S_m(\omega)d\omega$$

$$\iint_{-\infty}^{\infty} h(\xi)h(\xi')R_u(\xi - \xi')d\xi d\xi' = \frac{1}{2\pi}\int_{-\infty}^{\infty} S_u(\omega)d\omega$$

$$\cdot \int_{-\infty}^{\infty} e^{j\omega\xi}h(\xi)d\xi \int_{-\infty}^{\infty} e^{-j\omega\xi'}R(\xi')d\xi'$$

$$= \frac{1}{2\pi}\int_{-\infty}^{\infty} S_u(\omega)H(-\omega)H(\omega)d\omega$$

from which it follows that (7.42) can be written as

$$\overline{\epsilon^2} = \frac{1}{2\pi}\int_{-\infty}^{\infty} [S_m(\omega) - S_u(\omega)H(-\omega)H(\omega)]\,d\omega$$

Using relationship (7.41) the mean square error becomes

$$\overline{\epsilon^2} = \frac{1}{2\pi}\int_{-\infty}^{\infty} \frac{S_m(\omega)S_u(\omega) - S_{mu}(\omega)S_{mu}(-\omega)}{S_u(\omega)}\,d\omega \qquad (7.43)$$

where

$$S(\omega) = S(-\omega)$$

We now apply the formulation to the problem of simple filtering where

$$m(t) = s(t)$$

$$H = 1$$

We assume, for simplicity, there is no cross-correlation between the useful signal and noise, i.e.,

$$R_{sn}(\xi) = \overline{s(t)n(t-\xi)} = 0$$

The correlation functions R_{su} and R_u can be written as

$$R_{su}(\xi) = \overline{s(t)u(t-\xi)} = \overline{s(t)s(t-\xi)} + \overline{s(t)n(t-\xi)} = R_s(\xi)$$

$$R_u(\xi) = \overline{[s(t)+n(t)]\,[s(t-\xi)+n(t-\xi)]} = R_s(\xi) + R_n(\xi)$$

Consequently,

$$S_{su}(\omega) = S_s(\omega)$$

$$S_u(\omega) = S_s(\omega) + S_n(\omega)$$

and, from (7.41), the transfer function $H(\omega)$ is

$$H(\omega) = \frac{S_s(\omega)}{S_s(\omega) + S_n(\omega)} \tag{7.44}$$

The corresponding mean square error becomes, by equation (7.43),

$$\overline{\epsilon^2} = \frac{1}{2\pi}\int_{-\infty}^{\infty} \frac{S_s(\omega)S_n(\omega)}{S_s(\omega) + S_n(\omega)}\, d\omega \tag{7.45}$$

In the case where the spectral densities S_s and S_n do not overlap (Fig. 7.2a) we have, from equation (7.44),

$$H(\omega) = 1,\ S_s(\omega) \neq 0$$

$$H(\omega) = 0,\ S_s(\omega) = 0$$

From formula (7.45)

$$\overline{\epsilon^2} = 0$$

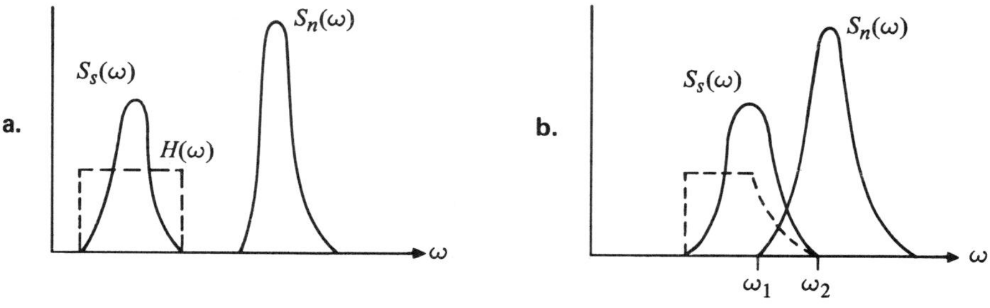

Figure 7.2 Filtering of random processes.

Hence, simple filtering, where the useful signal and the noise do not occupy the same frequency band, takes place without error. However, if S_s and S_n overlap (Fig. 7.2b) then error accompanies the filtering. The error is, in part, due to the noise in the frequency range $\omega_1 < \omega < \omega_2$ passing through the filter, and, in part, due to the distortion of the signal resulting from attenuation in the frequency range $\omega_1 < \omega < \omega_2$. As S_n gets larger and S_s gets smaller in the range $\omega_1 < \omega < \omega_2$ the less this range will be allowed to pass through the optimum filter.

We next examine the interesting case where

$$S_n(\omega) \gg S_s(\omega)$$

The spectral density of the noise is much larger than that of the useful signal. In this case the transfer function is approximated, from equation (7.44), as

$$H(\omega) = \frac{S_s(\omega)}{S_n(\omega)} \ll 1$$

The corresponding mean square error is, from (7.45),

$$\overline{\epsilon^2} = \frac{1}{2\pi} \int_{-\infty}^{\infty} S_s(\omega) d\omega = R_s(0)$$

$$= \overline{s^2(t)}$$

Thus, when the intensity of the input noise is much greater than the useful signal and occupies the same frequency range then the mean square error at the output of the optimum filter is equal to the mean square of the useful signal. This is to say that when the input noise is intense the intensity of $y(t)$ at the output is weak. In fact, we have, approximately,

$$y(t) = 0$$

$$\epsilon(t) = s(t)$$

since the magnitude of H is much less than unity.

7.5 SPECTRAL DENSITY OF SIGNAL AND NOISE

We assumed in the simple example of Section 7.4 that $R_{sn}(\xi) = 0$. However, in practice $R_{sn}(\xi) \neq 0$ and, therefore, $S_{sn}(\omega)$ is finite. Accordingly, one can ask what is the physical significance of $S_{sn}(\omega)$? From (7.36) the spectral density $S_{sn}(\omega)$ corresponding to the cross-correlation function $R_{sn}(\xi)$ is

$$S_{sn}(\omega) = S_{ns}(-\omega) \tag{7.46}$$

$$S_{sn}(-\omega) = S^{*}_{sn}(\omega) \tag{7.47}$$

For the process $u(t)$ we can write

$$\begin{aligned} R_u(\xi) &= \overline{[s(t) + n(t)]\,[s(t-\xi) + n(t-\xi)]} \\ &= R_s(\xi) + R_n(\xi) + R_{sn}(\xi) + R_{ns}(\xi) \end{aligned}$$

Therefore

$$\begin{aligned} S_u(\xi) &= S_s(\omega) + S_n(\omega) + S_{sn}(\omega) + S_{ns}(\omega) \\ &= S_s(\omega) + S_n(\omega) + 2Re[S_{sn}(\omega)] \end{aligned}$$

Thus, the spectral density of the input process consists of that which is signal, that which is noise, and an added real part $2Re[S_{sn}(\omega)]$ called the interference intensity caused by the statistical coupling between $s(t)$ and $n(t)$. The imaginary part of $S_{sn}(\omega)$, which identifies the phase of the statistical relationship between $s(t)$ and $n(t)$, has no explicit physical meaning.

7.6 POWER SPECTRAL DENSITY AND CORRELATION FUNCTION

Consider the stationary stochastic process of Appendix F. Let the process be ergodic where the sample function $u(t)$ contains all the characteristics of the process. For real $u(t)$ with Fourier transform $U(\omega)$ the energy E of $u(t)$ is

$$E = \int_{-\infty}^{\infty} u^2(t)dt = \frac{1}{2\pi}\int_{-\infty}^{\infty} |U(\omega)|^2 d\omega \tag{7.48}$$

The quantity $|U(\omega)|^2$ is the energy density spectrum whereas $u^2(t)$ is the instantaneous power associated with $u(t)$. The average power $\bar{P}$ of $u(t)$ is

$$\langle P \rangle = \bar{P} = \lim_{T \to \infty} \frac{1}{2T} \int_{-T}^{T} u^2(t)dt \tag{7.49}$$

It is possible from (7.48) and (7.49) that the energy of $u(t)$ can be infinite and the average power finite. If the energy of $u(t)$ is not finite it may not have a Fourier transform. Therefore it is convenient to define the truncated signal $u_T(t)$:

$$u_T(t) = \begin{cases} u(t) & |t| \leqslant T \\ 0 & |t| > T \end{cases}$$

The corresponding Fourier transform $U_T(\omega)$ is

$$U_T(\omega) = \int_{-\infty}^{\infty} u_T(t)e^{-j\omega t}dt = \int_{-T}^{T} u(t)e^{-j\omega t}dt \tag{7.50}$$

Thus, as $T \to \infty$ the truncated sample function $u_T(t)$ approaches $u(t)$. The average power $\bar{P}_T$ of $u(t)$ over the interval $(-T,T)$ may now be written as

$$\langle P_T \rangle = \bar{P}_T = \frac{1}{2T} \int_{-T}^{T} u^2(t)dt = \frac{1}{2T} \int_{-\infty}^{\infty} u_T^2(t)dt$$

$$= \frac{1}{2\pi} \int_{-\infty}^{\infty} |U(\omega)|^2/2T \, d\omega \tag{7.51}$$

The quantity $|U(\omega)|^2/2T$ is the power spectral density of $u(t)$ in the interval $(-T,T)$.

In dealing with a random process that is not ergodic, we may associate with each sample function $u(t)$ a truncated sample function $u_T(t)$, its Fourier transform $U_T(\omega)$ and its spectral density. The power spectral density $W(\omega)$ of the random process is defined as

$$W(\omega) = E\{|U_T(\omega)|^2/2T\}$$

$$= \lim_{T\to\infty} \frac{1}{2T} \overline{|U(\omega)|^2}$$

$$= \lim_{T\to\infty} \frac{1}{2T} \overline{U(\omega)\,U^*(\omega)} \tag{7.52}$$

i.e., it is the limit of the ensemble average of the power spectral densities of all of its truncated sample functions. Accordingly $U_T(\omega)$ is a random variable.

Equation (7.52) is not in convenient form to calculate power spectral densities. A more convenient form relates power spectral density to the correlation function. For stationary processes $W(\omega)$ is the Fourier transform of the correlation function. We see this by substituting the definition of $U_T(\omega)$ in (7.52), which gives

$$W(\omega) = \lim_{T\to\infty} \frac{1}{2T} \overline{U_T(\omega)\,U_T^*(\omega)}$$

$$= \lim_{T\to\infty} \frac{1}{2T} E\left\{\int_{-T}^{T} u(t)\,e^{-j\omega t}dt \int_{-T}^{T} u(t)\,e^{j\omega t}dt\right\}$$

$$= \lim_{T\to\infty} \frac{1}{2T} E\left\{\int_{-T}^{T} dt_1 \int_{-T}^{T} dt_2\, u(t_1)u(t_2)\,e^{-j\omega(t_1-t_2)}\right\} \tag{7.53}$$

The ensemble averaging of $u(t_1)$ and $u(t_2)$ is

$$\overline{u(t_1)u(t_2)} = R_u(t_1, t_2)$$

$$\overline{u(t-t_1)u(t-t_2)} = R_u(t_1 - t_2) \tag{7.54}$$

Substituting (7.54) in (7.53) we have

$$W(\omega) = \lim_{T\to\infty} \frac{1}{2T} \int_{-T}^{T} dt_1 \int_{-T}^{T} dt_2\, R_u(t_1 - t_2) e^{-j\omega(t_1-t_2)} \tag{7.55}$$

The double integral can be written as a single integral with the simple change of variables,

$$\xi = t_1 - t_2 \qquad \xi' = t_1 + t_2$$

$$d\xi = dt_1 \qquad d\xi' = dt_2$$

Accordingly, for any function $g(t_1 - t_2)$ one can write

$$\int_{-T}^{T} dt_1 \int_{-T}^{T} dt_2\, g(t_1 - t_2) = \frac{1}{2}\int_{-2T}^{2T} d\xi \int_{-2T+|\xi|}^{2T-|\xi|} g(\xi)d\xi'$$

$$= \int_{-2T}^{2T} (2T - |\xi|)g(\xi)d\xi$$

Using this expression in (7.55) we get

$$W(\omega) = \lim_{T\to\infty} \frac{1}{2T}\int_{-2T}^{2T} (2T - |\xi|)R_u(\xi)e^{-j\omega\xi}d\xi$$

$$= \lim_{T\to\infty} \frac{1}{2T}\int_{-2T}^{2T} \left(1 - \frac{|\xi|}{2T}\right) R_u(\xi)e^{-j\omega\xi}d\xi$$

$$= \int_{-\infty}^{\infty} R_u(\xi)e^{-j\omega\xi}d\xi \tag{7.56}$$

Thus, for stationary random processes the power spectral density is the Fourier transform of the correlation function.

In comparing (7.56) with (7.36) it is seen that from the definition of $S(\omega)$ we have

$$S(\omega) = W(\omega) \tag{7.57}$$

The inverse of (7.56) immediately gives

$$R_u(\xi) = \frac{1}{2\pi}\int_{-\infty}^{\infty} S(\omega)e^{j\omega\xi}d\omega$$

$$= \frac{1}{2\pi}\int_{-\infty}^{\infty} W(\omega)e^{j\omega\xi}d\xi \tag{7.58}$$

Relationships (7.56) and (7.58), between the power spectral density and the correlation function, are known as the *Wiener-Khintchine* equations.

7.7 CORRELATION TIME-PASS BAND PRODUCT

Using (7.36) and (7.37) we consider a few sample functions of $R(\xi)$ and corresponding $S(\omega)$:

	$R(\xi)$	$S(\omega)$
(a)	$R(0)e^{-\alpha\|\xi\|}$	$\dfrac{2\alpha R(0)}{\alpha^2 + \omega^2}$
(b)	$\dfrac{R(0)}{1 - \alpha^2\xi^2}$	$\dfrac{\pi R(0)}{\alpha} e^{-\omega/\xi}$
(c)	$\dfrac{R(0) \sin \omega_0\xi}{\omega_0\xi}$	$\begin{cases} \dfrac{\pi R(0)}{\omega_0} & -\omega_0 < \omega < \omega_0 \\ 0 & \text{otherwise} \end{cases}$

where α is a real parameter. In each case the functions $R(\xi)$ and $S(\omega)$ assume their maximum values when $\xi = 0$ and $\omega = 0$. We (loosely) define the correlation time $\Delta\xi$ as the time interval $-\Delta\xi < \xi < \Delta\xi$ within which the value of $R(\xi)$ is of the same order of magnitude as $R(0)$. Outside this time interval $R(\xi)$ is much less than $R(0)$. For the examples chosen $\Delta\xi$ is approximately equal to $1/\alpha$, i.e., $\Delta\xi \sim 1/\alpha$. Similarly, we define the spectral bandwidth $\Delta\omega$ as the frequency interval $-\Delta\omega < \omega < \Delta\omega$ within which $S(\omega)$ compares to $S(0)$. Outside this interval $S(\omega)$ is much less than $S(0)$. For the examples chosen $\Delta\omega \sim \alpha$. Hence the correlation time-bandwidth product

$$\Delta\xi\Delta\omega \sim 1 \tag{7.59}$$

By introducing a more exact definition of $\Delta\xi$ and $\Delta\omega$ we can derive a correspondingly more precise expression for $\Delta\xi\Delta\omega$. Let the spectral bandwidth $\Delta\omega$ be defined as

$$\int_{-\infty}^{\infty} S(\omega)d\omega = 2\int_0^{\infty} S(\omega)d\omega = 2S(0)\Delta\omega \tag{7.60}$$

Definition (7.60) specifies $\Delta\omega$ such that the curve $S(\omega)$ can be approximated by a rectangle of height $S(0)$ and width $2\Delta\omega$ as shown in Fig. 7.3). The area of the

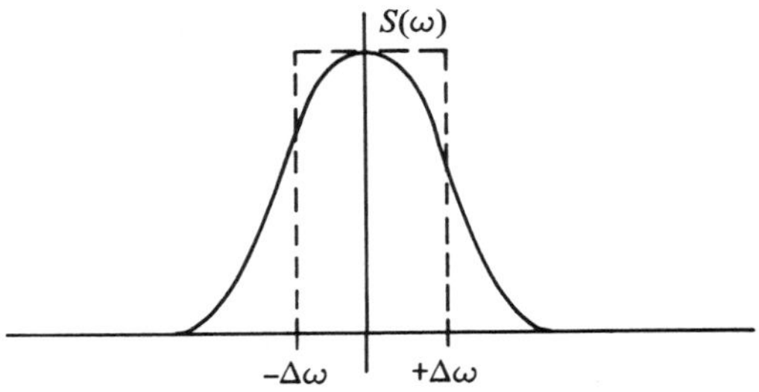

Figure 7.3 Spectral bandwidth.

rectangle is equal to the area under the curve $S(\omega)$. Similarly, the correlation time $\Delta\xi$ is defined as

$$\int_{-\infty}^{\infty} R(\xi)d\xi = 2\int_{0}^{\infty} R(\xi)d\xi = 2R(0)\Delta\xi \tag{7.61}$$

Using

$$S(0) = \int_{-\infty}^{\infty} R(\xi)d\xi \qquad R(0) = \frac{1}{2\pi}\int_{-\infty}^{\infty} S(\omega)d\omega$$

the exact relation for the correlation time-bandwidth product is

$$\Delta\xi\Delta\omega = \frac{\pi}{2} \tag{7.62}$$

Equations (7.59) and (7.62) apply to any pair of functions which are the Fourier transform of each other. Of particular interest is their application to the impulse response $h(\xi)$ and transfer function $H(\omega)$. The quantity $\Delta\omega$ defines the *pass-band* of the system. $\Delta\xi$ defines the correlation time or memory of the system. The system memory is taken to be the time interval during which the input appreciably influences the output. It symbolizes the time during which the system responds to a unit impulse.

7.8 TRANSFER FUNCTIONS AND LINEAR OPERATORS

The relationship between $s(t)$ and $m(t)$ was specified by (7.18) as

$$m(t) = H[s(t)]$$

where H is a linear operator. The Fourier transform of the impulse response [c.f. eq. (7.38)] established the transfer function $H(\omega)$ for the optimum filter as [c.f. eq. (7.41)]

$$H(\omega) = \frac{S_{mu}(\omega)}{S_u(\omega)}$$

Equation (7.41) is the direct result of the input-output relation

$$y(t) = \int_{-\infty}^{\infty} h(\xi)u(t-\xi)d\xi$$

We next investigate the relationship of H and $H(\omega)$. From (7.18) the integral form of the operator equation is

$$m(t) = \int_{-\infty}^{\infty} h(t')s(t-t')dt' \tag{7.63}$$

Previously we had

$$S_{mu}(\omega) = \int_{-\infty}^{\infty} e^{-j\omega\xi} R_{mu}(\xi)d\xi$$

$$= \int_{-\infty}^{\infty} e^{-j\omega\xi}\, \overline{m(t)u(t-\xi)}\, d\xi \tag{7.64}$$

Substituting $m(t)$ from (7.63) into (7.64) gives

$$S_{mu}(\omega) = \int_{-\infty}^{\infty} e^{-j\omega\xi} d\xi \int_{-\infty}^{\infty} h(t')s(t-t')u(t-\xi)dt'$$

$$= \int_{-\infty}^{\infty} e^{-j\omega\xi} d\xi \int_{-\infty}^{\infty} h(t')R_{su}(\xi-t')dt' \tag{7.65}$$

By changing the variable of integration, such that $\xi = t + t'$, equation (7.65) can be written as

$$S_{mu}(\omega) = \int_{-\infty}^{\infty} e^{-j\omega t'} h(t')dt' \int_{-\infty}^{\infty} e^{-j\omega t} R_{su}(t)dt$$

$$= H(\omega)S_{su}(\omega) \tag{7.66}$$

Thus the transfer function as derived from the operator relation (7.63) is the ratio of the spectral densities $S_{mu}(\omega)$ and $S_{su}(\omega)$, where we have specified the general form of $H(\omega)$ as

$$H(\omega) = \int_{-\infty}^{\infty} e^{-j\omega t} h(t)dt \tag{7.67}$$

Obviously, not all linear operators can be written in the form of (7.63). In these cases the transfer function $H(\omega)$ is *defined* to be

$$H(\omega) = \frac{S_{mu}(\omega)}{S_{su}(\omega)}$$

A few simple examples illustrating this point are as follows: For simple filtering H is the unit operator, i.e.,

$$m(t) = H[s(t)] = s(t)$$

and $S_{mu}(\omega) = S_{su}(\omega)$. Thus the transfer function

$$H(\omega) = 1$$

For time-shift operators where

$$m(t) = s(t + \Delta)$$

it is readily shown that

$$H(\omega) = e^{-j\omega\Delta}$$

The differentiation operator

$$m(t) = \frac{d}{dt}\{s(t)\}$$

gives

$$H(\omega) = j\omega$$

whereas the integral operator

$$m(t) = \int_t dt\,\{s(t)\}$$

$$= \left(\frac{d}{dt}\right)^{-1}\{s(t)\}$$

gives

$$H(\omega) = \frac{1}{j\omega}$$

Each of the above identities can be proven by forming $S_{mu}(\omega)$ and $S_{su}(\omega)$ and applying formula (7.66).

Generalizing the above equations it can be said that if the transfer function $H(\omega)$ is a polynomial

$$H(\omega) = \sum_{\ell=0}^{n} C_\ell (j\omega)^\ell \tag{7.68}$$

where the coefficients C_ℓ are constants, then the corresponding operator can be written as

$$H = \sum_{\ell=0}^{n} C_\ell \left(\frac{d}{dt}\right)^\ell \tag{7.69}$$

Formula (7.69) applies only when (7.66) can be written as the ratio of two polynomials

$$H(\omega) = \frac{P_b(\omega)}{P_a(\omega)} \tag{7.70}$$

where $P_a(\omega)$ and $P_b(\omega)$ are polynomials of degree a and b, respectively, and $a > b$. If $a \leqslant b$ the numerator is divided by the denominator giving

$$H(\omega) = H_0(\omega) + H_1(\omega) \tag{7.71}$$

where

$$H_0(\omega) = \sum_{\ell=0}^{b-a} C_\ell (j\omega)^\ell \tag{7.72}$$

and the remainder

$$H_1(\omega) = \int_{-\infty}^{\infty} h_1(\xi) u(t-\xi) d\xi \tag{7.73}$$

7.9 MONOCHROMATIC SIGNALS

The classical form of an amplitude modulated signal generated by an harmonic oscillator is

$$\begin{aligned} s(t) &= a \cos \omega_0 t + b \sin \omega_0 t \\ &= e \cos(\omega_0 t + \varphi) \end{aligned} \tag{7.74}$$

where

$$a = e \cos \varphi \qquad b = e \sin \varphi$$

Let the modulation parameters a and b, or e and ϕ, be random variables which are independent of time and satisfy the conditions

$$\bar{a} = \bar{b} = 0 \qquad \overline{a^2} = \overline{b^2} \qquad \overline{ab} = 0 \tag{7.75}$$

Then waveform (7.74) can be considered as a random process. In order for this process to qualify as a quasi-monochromatic signal the bandwidth $\Delta\omega$ of the modulation must be very small compared to the carrier frequency ω_0, i.e.,

$$\Delta\omega \ll \omega_0 \tag{7.76}$$

In accordance with conditions (7.75) the stationary random functions $a(t)$ and $b(t)$ have the following statistical properties

$$\overline{a(t)} = \overline{b(t)} = 0 \tag{7.77}$$

$$\overline{a(t)a(t-\xi)} = \overline{b(t)b(t-\xi)} = c^2 r(\xi) \tag{7.78}$$

$$\overline{a(t)b(t-\xi)} = \overline{b(t)a(t-\xi)} = 0 \tag{7.79}$$

where $c^2r(\xi)$ is the autocorrelation function for processes $a(t)$ and $b(t)$. By property (7.79) there is no cross-correlation between the two processes. The constant c^2 is determined by

$$c^2 = \overline{a^2} = \overline{b^2} \tag{7.80}$$

i.e., it is the mean square value of a and b, so that

$$r(0) = 1 \tag{7.81}$$

Correlation functions satisfying conditions (7.81) are said to be *normalized*.

Calculating the correlation function for $s(t)$ we have

$$\begin{aligned} R_s(\xi) &= \overline{[a(t)\cos\omega_0 t + b(t)\sin\omega_0 t][a(t-\xi)\cos\omega_0(t-\xi) + b(t-\xi)\sin\omega_0(t-\xi)]} \\ &= c^2r(\xi)[\cos\omega_0 t\cos\omega_0(t-\xi) + \sin\omega_0 t\sin\omega_0(t-\xi)] \\ &= c^2r(\xi)\cos\omega_0\xi \end{aligned} \tag{7.82}$$

On normalizing we have for $\xi = 0$

$$R_s(0) = c^2 = \overline{s^2(t)} \tag{7.83}$$

The spectral density for the normalized correlation function $r(\xi)$ is

$$s(\omega) = \int_{-\infty}^{\infty} e^{-j\omega\xi}r(\xi)d\xi \tag{7.84}$$

Its corresponding inverse becomes

$$r(\xi) = \frac{1}{2\pi}\int_{-\infty}^{\infty} e^{j\omega\xi}s(\omega)d\omega \tag{7.85}$$

Thus the spectral density $S_s(\omega)$ of the quasi-monochromatic signal $s(t)$ can be written as

$$S_s(\omega) = \int_{-\infty}^{\infty} e^{-j\omega\xi}R_s(\xi)d\xi = c^2\int_{-\infty}^{\infty} e^{-j\omega\xi}r(\xi)\cos\omega_0\xi\, d\xi$$

$$= \frac{c^2}{2}\left[\int_{-\infty}^{\infty} e^{-j(\omega-\omega_0)}r(\xi)d\xi + \int_{-\infty}^{\infty} e^{-j(\omega+\omega_0)}r(\xi)d\xi\right]$$

$$= \frac{c^2}{2}[s(\omega - \omega_0) + s(\omega + \omega_0)] \tag{7.86}$$

We assume that $s(\omega)$ is bell-shaped with its maximum at $\omega = 0$. Therefore from (7.86) and (7.76) $S_s(\omega)$ represents two non-overlapping bell-shaped curves centered approximately at $-\omega_0$ and $+\omega_0$. From (7.62) where

$$\Delta\xi\Delta\omega = \frac{\pi}{2}$$

inequality (7.76) implies that

$$\omega_0\Delta\xi \gg 1 \tag{7.87}$$

Since $\Delta\xi$ defines the variation with time of the random functions $a(t)$ and $b(t)$ condition (7.87) specifies that $\cos \omega_0 t$ and $\sin \omega_0 t$ undergo many oscillations in time before $a(t)$ and $b(t)$ change appreciably.

For the ensuing calculations we *choose* the form of the normalized correlation function as

$$r(\xi) = e^{-\alpha|\xi|} \tag{7.88}$$

where α determines the spectral bandwidth. The corresponding Fourier transform gives

$$s(\omega) = \frac{2\alpha}{\alpha^2 + \omega^2} \tag{7.89}$$

From (7.81) and (7.85) we have

$$\frac{1}{2\pi}\int_{-\infty}^{\infty} s(\omega)d\omega = 1$$

which when combined with (7.60) gives

$$\Delta\omega = \frac{\pi}{2}\alpha \tag{7.90}$$

We now examine the optimum filter which separates the quasi-monochromatic signal from so-called "white noise," i.e., noise of constant spectral density. We have

$$S_n(\omega) = S_n = \text{const.} \tag{7.91}$$

For simple filtering where there is no correlation between signal and noise, i.e., where

$$\overline{s(t)n(t-\xi)} = 0$$

the transfer function is, by (7.44),

$$H(\omega) = \frac{S_s(\omega)}{S_s(\omega) + S_n(\omega)}$$

and

$$h(\xi) = \frac{1}{2\pi}\int_{-\infty}^{\infty} \frac{e^{j\omega\xi}\,d\xi}{1 + [S_n(\omega)/S_s(\omega)]} \tag{7.92}$$

Our solution to (7.92) will be confined to positive values of ξ, since $h(\xi)$ is an even function. Representing the denominator of (7.92) as $D(\omega)$ we look for the roots of the equation $D(\omega) = 0$. For $\omega \sim \omega_0$ we have approximately, from (7.86),

$$S_s(\omega) \sim \frac{c^2}{2}s(\omega - \omega_0) = \frac{c^2\alpha}{(\omega - \omega_0)^2 + \alpha^2}$$

where the term $s(\omega - \omega_0) \sim s(2\omega_0)$ is neglected. Thus, the denominator

$$D(\omega) = 1 + S_n \frac{(\omega - \omega_0)^2 + \alpha^2}{c^2\alpha}$$

has the roots

$$\omega = \omega_0 \pm j\theta$$

where

$$\theta = \sqrt{\alpha^2 + (c^2\alpha/S_n)}$$

Similarly, two other zeros of $D(\omega)$ can be found for $\omega \sim -\omega_0$. At the poles $\omega = \pm\omega_0 + j\theta$ of the integrand in (7.92) the derivative $dD/d\omega$ is

$$\frac{dD}{d\omega} = j\frac{2\theta S_n}{c^2\alpha}$$

where we have asserted that only the poles in the upper half-plane are important. The contour of integration is closed in the upper half-plane and the integral is reduced to a sum of residues. The final expression for $h(\xi)$ becomes

$$h(\xi) = \frac{\alpha^2\rho}{\theta} e^{-\theta|\xi|} \cos \omega_0\xi \tag{7.93}$$

where

$$\rho = \frac{c^2}{\alpha S_n} = \frac{\pi}{2} \frac{\overline{s^2(t)}}{S_n \Delta\omega} \tag{7.94}$$

The dimensionless parameter ρ represents the *signal-to-noise power ratio.* It depicts the extent to which the signal is stronger (or weaker) than the noise. From the above the parameter θ can be written in terms of α as

$$\theta = \alpha\sqrt{1+\rho} \tag{7.95}$$

According to (7.83) c^2 represents the average power of the quasi-monochormatic signal, and by (7.90) $S_n\alpha$ is the noise power in the band occupied by the signal.

For small signal-to-noise ratio $\rho \ll 1$. Therefore equation (7.95) implies

$$\theta = \alpha$$

Hence, from (7.93), (7.94) and (7.88) the impulse response is

$$h(\xi) = \alpha\rho e^{-\alpha|\xi|} \cos \omega_0\xi$$

$$= \frac{R_s(\xi)}{S_n} \tag{7.96}$$

and fron (7.44) the transfer function reduces to

$$H(\omega) = \frac{S_s(\omega)}{S_n} \tag{7.97}$$

which is in agreement with our discussion in Section 7.4.

The pass-band established by the roots of $D(\omega)$ include the frequency intervals $\omega_0 \pm \Delta\omega$ and $-\omega_0 \pm \Delta\omega$ where, from (7.90) and (7.93), the spectral bandwidth $\Delta\omega$ is determined by the parameter θ:

$$\Delta\omega = \frac{\pi}{2}\theta$$

$$= \frac{\pi}{2}\alpha\sqrt{1+\rho} \qquad (7.98)$$

We see from (7.98) that as the noise power S_n decreases the pass-band $\Delta\omega$ increases. This leads to a shortening of the (memory) time $1/\theta$ required for the signal to be extracted from the noise, which agrees with our earlier conclusion that for simple filtering in the absence of noise no time is required to duplicate the signal. However, the extraction of signals in the presence of noise must be done in a time less than the correlation time of the signal, in which case we always have $\theta > \alpha$.

The mean-square filtering error for the quasi-monochromatic signal is

$$\overline{\epsilon^2} = \overline{s^2} - \frac{1}{2\pi}\int_{-\infty}^{\infty} |H(\omega)|^2 S_u(\omega)\,d\omega$$

$$= c^2 - \frac{1}{2\pi}\int_{-\infty}^{\infty} \frac{S_s^2(\omega)}{S_s(\omega) + S_n(\omega)}\,d\omega$$

$$= c^2 - \frac{1}{2\pi}\int_{-\infty}^{\infty} \frac{S_s(\omega)}{1 + [S_n/S_s(\omega)]}\,d\omega \qquad (7.99)$$

The integral of (7.99) is evaluated by calculating the residues associated with the zeros of the denominator ($\omega = \pm\omega_0 + j\theta$) and the residues associated with the poles of the numerator ($\omega = \pm\omega_0 + j\alpha$). Equation (7.99) becomes

$$\overline{\epsilon^2} = -\frac{S_s(\omega_0 + j\theta) + S_s(-\omega_0 + j\theta)}{2S_n\theta/c^2\alpha}$$

$$= \frac{c^2\alpha}{\theta} = \frac{c^2}{\sqrt{1+\rho}} \qquad (7.100)$$

Consider the situation where the signal is "highly" monochromatic, i.e., the parameter α approaches zero. Therefore, by (7.94) ρ then increases and by

(7.100) $\overline{\epsilon^2}$ decreases. Effectively, the filter pass-band $\Delta\omega$ becomes narrower and its correlation time $\Delta t \sim 1/\theta$ becomes larger thereby making more effective use of the signal. For strictly monochromatic signals $\alpha = \theta = \omega = 0$ and $\overline{\epsilon^2} = 0$. In this case the infinite working time of the filter produces an infinitely small (or zero) error. Clearly, for a finite working time the error is greater than zero.

7.10 SIGNALS OF KNOWN FORM IN NOISE

In our previous discussions we treated the signal and noise as random processes. In the disciplines of radar and radio communications the signal is usually of known form and is not regarded as a random process. Instead, it can be regarded as a known function with several unknown parameters, such as amplitude or phase. In such cases filtering must, (a) make the most reliable observation of the useful signal, and (b) make the most precise measurement of the signal's unknown parameters. Consequently, a performance metric of a filter extracting signals from noise can be the signal-to-noise ratio at the output.

Assuming the signal $s(t)$ has a well-defined form we have from (7.16)

$$u(t) = s(t) + n(t)$$

where $n(t)$ is a random process. The corresponding output is

$$\begin{aligned} y(t) &= H[u(t)] \\ &= H[s(t)] + H[n(t)] \\ &= \sigma(t) + \eta(t) \end{aligned} \tag{7.101}$$

where $\sigma(t)$ and $\eta(t)$ are the results of passing the useful signal and noise, respectively, through the filter. For $s(t)$ we can write its Fourier transform and inverse, respectively, as

$$S(\omega) = \int_{-\infty}^{\infty} e^{-j\omega t} s(t)\, dt$$

$$s(t) = \frac{1}{2\pi} \int_{-\infty}^{\infty} e^{j\omega t} S(\omega)\, d\omega$$

Thus, the useful signal at the filter output is

$$\sigma(t) = \frac{1}{2\pi}\int_{-\infty}^{\infty} e^{j\omega t}H(\omega)S(\omega)\,d\omega \tag{7.102}$$

Treating the noise as a random process we can write for $\overline{n^2(t)}$ [c.f. equation (7.48)], the average noise power at the input,

$$\overline{n^2(t)} = \frac{1}{2\pi}\int_{-\infty}^{\infty} S_n(\omega)\,d\omega \tag{7.103}$$

The noise at the output of the filter becomes

$$\overline{\eta^2(t)} = \frac{1}{2\pi}\int_{-\infty}^{\infty} |H(\omega)|^2 S_n(\omega)\,d\omega \tag{7.104}$$

Accordingly, we *define* the *signal-to-noise power ratio* ρ as

$$\rho = \frac{[\sigma(t_0)]^2}{\overline{\eta^2}} \tag{7.105}$$

where $\sigma(t_0)$ is the signal value at a specified time t_0. From (7.102) and (7.104) the signal-to-noise power ratio can be written as

$$\rho = \frac{1}{2\pi}\,\frac{\left|\int_{-\infty}^{\infty} e^{j\omega t_0}H(\omega)S(\omega)\,d\omega\right|^2}{\int_{-\infty}^{\infty} |H(\omega)|^2 S_n(\omega)\,d\omega} \tag{7.106}$$

We now look for a filter which gives the largest value of ρ. Our analytical procedure will be to use the relation

$$y(t_0) = \sigma(t_0) + \eta(t_0) \tag{7.107}$$

to decide whether or not a signal is present, and make

$$|\sigma(t_0)| > \sqrt{\overline{\eta^2(t_0)}} \tag{7.107a}$$

by as much as possible. Applying the Schwarz inequality

$$\left| \int_{-\infty}^{\infty} e^{j\omega t_0} H(\omega) S(\omega) \, d\omega \right|^2 \leqslant \int_{-\infty}^{\infty} |H(\omega)|^2 S_n(\omega) \, d\omega \int_{-\infty}^{\infty} \frac{|S(\omega)|^2}{S_n(\omega)} \, d\omega \qquad (7.108)$$

to (7.106) the upper bound of the signal-to-noise ratio can be established as

$$\rho \leqslant \frac{1}{2\pi} \int_{-\infty}^{\infty} \frac{|S(\omega)|^2}{S_n(\omega)} \, d\omega \qquad (7.109)$$

If we let the transfer function $H(\omega)$ have the form

$$H(\omega) = k e^{-j\omega t_0} \frac{S^*(\omega)}{S_n(\omega)} \qquad (7.110)$$

where k is an arbitrary constant, ρ reaches its maximum in (7.109). We have

$$\rho = \frac{1}{2\pi} \int_{-\infty}^{\infty} \frac{|S(\omega)|^2}{S_n(\omega)} \, d\omega \qquad (7.111)$$

Thus, a linear filter with the transfer function (7.110) is the best filter among linear filters. Further, if the noise $n(t)$ is a normal random process (i.e., Gaussian) the filter (7.110) is an absolute optimum. Physically, (7.110) is interpreted to mean the larger the amplitude spectrum of the useful signal and the smaller the power spectrum of the noise in the frequency interval $(\omega, \omega + d\omega)$, the more the optimum filter will pass those frequencies. Also, from (7.111) it is apparent that the greater the displacement of frequency spectra of the useful signal and noise the greater the signal-to-noise ratio at the filter output.

In using filter (7.110) to detect signals of known form it is only necessary to know the value of (7.107), i.e., the value of the output function at time t_0. According to (7.102) and (7.110) we have

$$\sigma(t) = \frac{k}{2\pi} \int_{-\infty}^{\infty} e^{j\omega(t-t_0)} \frac{|S(\omega)|^2}{S_n} \, d\omega \qquad (7.112)$$

from which it follows that

$$|\sigma(t)| \leqslant |\sigma(t_0)| = |k| \rho \qquad (7.113)$$

By setting $k = 1$ in (7.110) we see that, according to (7.111) and (7.112),

$$\sigma(t_0) = \rho \tag{7.114}$$

and according to (7.105)

$$\overline{\eta^2(t)} = \rho \tag{7.114a}$$

i.e., the signal-to-noise ratio gives simultaneously the useful signal and the noise power at the filter output. In this case it can be shown that at the filter output the useful signal is related to the correlation function of the noise according to the formula

$$\sigma(t) = R_\eta(t_0 - t) \tag{7.115}$$

7.11 THE MATCHED FILTER

In the case where the noise spectrum is uniformly distributed over the useful frequency band we have the condition called "white noise." For this condition

$$S(\omega) = S_n = \text{const.} \tag{7.116}$$

From (7.110) the transfer function is

$$H(\omega) = e^{-j\omega t_0} S^*(\omega) \tag{7.117}$$

where arbitrarily we have chosen

$$k = S_n$$

Thus, for signals of known waveform wherein the noise is treated as a constant, the filter can be matched to the waveform (or its conjugate). Therefore, it is usually called the matched filter (or conjugate filter).

Since the signal $s(t)$ is real the conjugate of its Fourier transform can be written as

$$S^*(\omega) = S(-\omega) \tag{7.118}$$

The signal at the output of the filter is, according to (7.102), (7.117) and (7.118),

$$\sigma(t) = \frac{1}{2\pi}\int_{-\infty}^{\infty} e^{j\omega t} H(\omega)S(\omega)\,d\omega$$

$$= \frac{1}{2\pi}\int_{-\infty}^{\infty} e^{j\omega(t-t_0)} S(-\omega)d\omega \int_{-\infty}^{\infty} e^{-j\omega t'} s(t')\,dt' \qquad (7.119)$$

which, upon changing the order of integration, gives

$$\sigma(t) = \frac{1}{2\pi}\int_{-\infty}^{\infty} s(t')dt' \int_{-\infty}^{\infty} e^{j\omega(t-t_0-t')} S(-\omega)\,d\omega$$

$$= \frac{1}{2\pi}\int_{-\infty}^{\infty} s(t')dt' \int_{-\infty}^{\infty} e^{j\omega(t'-t+t_0)} S(\omega)\,d\omega$$

$$= \int_{-\infty}^{\infty} s(t')\, s(t'-t+t_0)\,dt' \qquad (7.120)$$

Clearly (7.120) has the form

$$\int_{-\infty}^{\infty} s(t)s(t-\xi)d\xi$$

Earlier [c.t. eq. (F.17)] we defined the time-average or integral form of the autocorrelation function $R_s(\xi)$ as

$$R_s(\xi) = \int_{-\infty}^{\infty} s(t)s(t-\xi)d\xi \qquad (7.121)$$

The integral form of $R_s(\xi)$ differs from definition (7.10). Instead of taking the ensemble average $\overline{s(t)s(t-\xi)}$ for a large number of identical experiments, we observe a single sample function which contains all the statistics of the process.

From (7.121) we have for an ergodic process

$$R_s(0) = \int_{-\infty}^{\infty} s^2(t)dt = E \tag{7.122}$$

which is in keeping with (7.48).

Substituting (7.121) into (7.120) the signal at the filter output is

$$\sigma(t) = R_s(t - t_0) \tag{7.123}$$

i.e., the matched filter is a *correlator*; its response to the useful signal is its correlation function. At $t = t_0$ the output becomes

$$\sigma(t_0) = R_s(0) = E \tag{7.124}$$

Khintchine's theorem states that for any random process $u(t)$

$$S_u(\omega) \geqslant 0$$

Therefore, the following inequality holds:

$$|R(\xi)| = \frac{1}{2\pi}\left|\int_{-\infty}^{\infty} S_u(\omega)e^{j\omega\xi}d\xi\right| \leqslant \frac{1}{2\pi}\int_{-\infty}^{\infty} S_u(\omega)d\omega$$

Thus, it follows that

$$R_s(0) \geqslant |R_s(\xi)| \qquad \sigma(t_0) \geqslant |\sigma(t)| \tag{7.125}$$

$\sigma(t_0)$ is the maximum value of the useful signal at the filter output; this maximum value is the total energy of the signal.

From (7.111) and (7.116) the signal-to-noise ratio is

$$\rho = \frac{1}{2\pi S_n}\int_{-\infty}^{\infty} |S(\omega)|^2 d\omega \tag{7.126}$$

And, from (7.118) and (7.122) we have the identity

$$\frac{1}{2\pi}\int_{-\infty}^{\infty} |S(\omega)|^2 d\omega = \int_{-\infty}^{\infty} s^2(t)dt = E$$

Thus, the signal-to-noise ratio takes on the simple form

$$\rho = \frac{E}{S_n} \tag{7.127}$$

i.e., it is determined by the energy content of the signal and the noise spectral density. For situations where the background noise is "constant" the only way to improve detecting a signal of known form is to increase its total energy.

According to (7.39) the impulse response $h(t)$ for the matched filter can be written as

$$\begin{aligned} h(t) &= \frac{1}{2\pi}\int_{-\infty}^{\infty} e^{j\omega t} H(\omega)\, d\omega \\ &= \frac{1}{2\pi}\int_{-\infty}^{\infty} e^{j\omega(t-t_0)} S(-\omega)\, d\omega \\ &= s(t - t_0) \end{aligned} \tag{7.128}$$

Thus, the impulse response of the best filter in white noise is the mirror image of the signal, delayed by t_0 seconds. Therefore, by (7.23) the output of the matched filter is of the form

$$\sigma(t) = \int_{-\infty}^{\infty} s(t' - t + t_0) u(t') dt'$$

The matched filter forms the *cross correlation* between the useful signal $s(t)$ and the input $u(t)$.

To appraise the operation of the matched filter we assume the useful signal has a sinusoidal waveform and contains energy E. We specify the transfer function as "rectangular" having the following characteristics:

$$H(\omega) = 1 \qquad \text{for } -\omega - \Delta\omega < \omega < -\omega + \Delta\omega$$

$$H(\omega) = 0 \qquad \text{for } \omega_0 - \Delta\omega < \omega < \omega_0 + \Delta\omega \tag{7.129}$$

Our objective is to compare the signal-to-noise ratio of the "mismatched" rectangular filter with that of the matched filter giving

$$\frac{\rho}{\rho_0} = \frac{\alpha \dfrac{E}{S_n}}{\dfrac{E}{S_n}} = \alpha$$

where ρ_0 is the signal-to-noise ratio of the matched filter. The exercise below will be to determine the value of α.

We specify for the sinusoidal pulse under consideration the appropriate form

$$\begin{aligned} s(t) &= A \cos \omega_0 t \qquad -\frac{T_0}{2} < t < \frac{T_0}{2} \\ &= 0 \qquad t < \frac{-T_0}{2},\ t > \frac{T_0}{2} \end{aligned} \tag{7.130}$$

where ω_0 is the carrier frequency, A is the pulse amplitude and T_0 is the pulse duration (or period). One can write for $S(\omega)$

$$\begin{aligned} S(\omega) &= \int_{-\infty}^{\infty} e^{-j\omega t} s(t)\,dt \\ &= A \int_{-T_0/2}^{T_0/2} e^{-j\omega t} \cos \omega_0 t\,dt \\ &= A \left[\frac{\sin(\omega - \omega_0)T_0/2}{\omega - \omega_0} + \frac{\sin(\omega + \omega_0)T_0/2}{\omega + \omega_0} \right] \end{aligned} \tag{7.131}$$

For the conditions where $\omega_0 T \gg 1$ equation (7.131) becomes

$$S(\omega) = A \frac{\sin(\omega - \omega_0)T_0/2}{\omega - \omega_0} \tag{7.132}$$

From (7.102) the filter output for the simple filtering specified by (7.129) is

$$\sigma(t) = \frac{1}{2\pi} \int_{-\omega_0 - \Delta\omega}^{-\omega_0 + \Delta\omega} e^{j\omega t} S(\omega)\,d\omega + \frac{1}{2\pi} \int_{\omega_0 - \Delta\omega}^{\omega_0 + \Delta\omega} e^{j\omega t} S(\omega)\,d\omega$$

$$= \frac{1}{\pi}\int_{\omega_0-\Delta\omega}^{\omega_0+\Delta\omega} S(\omega)\cos\omega t\, d\omega \tag{7.133}$$

By substituting (7.132) into (7.133) we have

$$\sigma(t) = \frac{A}{\pi}\int_{\omega_0-\Delta\omega}^{\omega_0+\Delta\omega} \cos\omega t\, \frac{\sin(\omega-\omega_0)T_0/2}{\omega-\omega_0}\, d\omega \tag{7.134}$$

If we impose the condition

$$\frac{\Delta\omega T_0}{2} \leqslant \pi \tag{7.135}$$

then

$$\frac{\sin(\omega-\omega_0)T_0/2}{\omega-\omega_0}$$

is positive over the integration interval. Therefore

$$|\sigma(t)| \leqslant \frac{A}{\pi}\int_{\omega_0-\Delta\omega}^{\omega_0+\Delta\omega} |\cos\omega t|\, \frac{\sin(\omega-\omega_0)T_0/2}{\omega-\omega_0}\, d\omega \leqslant \sigma(0) \tag{7.136}$$

where

$$\sigma(0) = \frac{A}{\pi}\int_{\omega_0-\Delta\omega}^{\omega_0+\Delta\omega} \frac{\sin(\omega-\omega_0)T_0/2}{\omega-\omega_0}\, d\omega = \frac{2A}{\pi}\int_0^{\nu} \frac{\sin x}{x}\, dx \tag{7.137}$$

$$x = (\omega-\omega_0)T_0/2; \qquad \nu = \Delta\omega T_0/2 \tag{7.138}$$

From (7.104) and condition (7.116) the noise intensity at the filter output is

$$\overline{\eta^2} = \frac{1}{\pi}\int_{\omega_0-\Delta\omega}^{\omega_0+\Delta\omega} S_n(\omega)d\omega$$

$$= \frac{2S_n}{\pi}\Delta\omega \tag{7.139}$$

Therefore the signal-to-noise ratio for simple filtering is

$$\rho = \frac{\sigma^2(0)}{\overline{\eta^2}} = \frac{2A^2X^2}{\pi S_n \Delta\omega} \tag{7.140}$$

where

$$X = \int_0^{\nu} \frac{\sin x \, dx}{x} \tag{7.141}$$

The energy of the rectangular pulse is, from (7.122),

$$E = \int_{-T_0/2}^{T_0/2} s^2(t)\,dt = A^2 \frac{T_0}{2} \tag{7.142}$$

which, when substituted into the signal-to-noise formula (7.140), gives

$$\rho = \frac{4EX^2}{\pi T_0 S_n \Delta\omega} = \frac{2EX^2}{\pi \nu S_n} \tag{7.143}$$

Forming the ratio ρ/ρ_0, where ρ_0 is the signal-to-noise ratio of the matched filter [c.f. eq. (7.127)], we have

$$\frac{\rho}{\rho_0} = \frac{2X^2}{\pi\nu} \tag{7.144}$$

The ratio ρ/ρ_0 depends on the parameter ν to the extent that the ratio is a maximum for

$$\nu \sim 2.15 \tag{7.145}$$

or

$$\omega \sim \frac{4.3}{T_0} \tag{7.146}$$

Thus

$$\frac{\rho}{\rho_0} \sim 0.825 \tag{7.147}$$

i.e., ρ is approximately one decibel less than ρ_0. Hence, the matched filter gives a larger signal to noise ratio because its transfer function fits the signal spectrum from the standpoint of both its bandwidth and shape (amplitude).

8
Quantized Systems-Perturbation Theory and State Transitions

8.1 INTRODUCTION

As an introduction to some of the methods of statistical analysis we will explore in a preliminary fashion some of the probabilistic methods of quantum theory, particularly as the theory describes the probability of state transitions resulting from small energy perturbations. The theory has experienced a high degree of success in physics and chemistry, and is introduced here primarily as a method for analysis. In the quantum picture the system is analyzed on the basis of its energy content, which is quantized. Associated with each of the quantized energy levels are the system states. The input to the system represents a change (perturbation) in the total energy of the system. This change results in the system transitioning to various different *allowable* energy states. The probability of a (state) transition occurring is directly related to the perturbing energy and the system's allowable quantized energy levels.

There exists, however, a fundamental difference, between the quantized system and the linear system. This difference arises from the theoretical concepts governing the respective theories and is manifested in the system dynamic equations. It was seen earlier that the dynamic state equation for the linear system is

$$|\dot{\mathbf{x}}\rangle = \mathbf{A}|\mathbf{x}\rangle + \mathbf{B}|\mathbf{u}\rangle$$

which is an outgrowth of the idea of ordered input-output pairs and techniques to attach a (state) "label" to the pairs. The "motion" of a quantized system is governed by the Schrodinger equation

$$j\hbar|\dot{\mathbf{x}}\rangle = \mathbf{H}|\mathbf{x}\rangle$$

which, as will be seen below, is one of the fundamental postulates of quantum theory.

8.2 FUNDAMENTAL POSTULATES OF QUANTUM THEORY

The intended goal here is to apply the methods of quantum theory to achieve simple rules in describing system *state transitions*. In describing these methods we first examine those fundamental postulates[1] which importantly serve as the basis for the theory. They are as follows:

POSTULATE A. *To every measurable, real quantity F there is associated an (Hermitian) operator* **F** *called an observable.*

For a Hermitian operator to be an observable it is required that the eigenvectors associated with the operator span the entire space. Thus, the eigenvectors $|\mathbf{x}_i\rangle$ of the observable can serve as a complete set of orthonormal base vectors where

$$\langle \mathbf{x}_i | \mathbf{x}_j \rangle = \delta_{ij}$$

Any finite state vector $|\mathbf{x}\rangle$ in this space can be described in terms of the eigenvectors:

$$|\mathbf{x}\rangle = \sum_i c_i |\mathbf{x}_i\rangle \tag{8.1}$$

where the c_i are the expansion coefficients.

1. See, for example, Grossman, L. M., *Thermodynamics and Statistical Mechanics*, McGraw-Hill, New York, 1969, pp. 77-87.

POSTULATE B. *The only possible values of a measurement of the measurable quantity F are the eigenvalues* λ *of the operator* $\mathbf{F}$.

From Postulate A the $|\mathbf{x}_i\rangle$ are the eigenvectors of $\mathbf{F}$; i.e.,

$$\mathbf{F}|\mathbf{x}_i\rangle = \lambda_i|\mathbf{x}_i\rangle \tag{8.2}$$

The scalar product formed by (8.1) and the ith eigenvector identifies the ith coefficient c_i as

$$c_i = \langle\mathbf{x}_i|\mathbf{x}\rangle \tag{8.3}$$

The probability that the eigenvalue λ_i of $\mathbf{F}$ will be measured in state $|\mathbf{x}\rangle$ is $c_i^* c_i$, where $c_i|\mathbf{x}_i\rangle$ is the projection of $|\mathbf{x}\rangle$ along the normalized basis eigenvector $|\mathbf{x}_i\rangle$.

POSTULATE C. *If a system is in a state represented by the vector* $|\mathbf{x}\rangle$ *the expected mean value resulting from a number of measurements of the measurable quantity F, whose observable is* $\mathbf{F}$, *is*

$$\langle F\rangle = \frac{\langle\mathbf{x}|\mathbf{F}|\mathbf{x}\rangle}{\langle\mathbf{x}|\mathbf{x}\rangle} \tag{8.4}$$

where the double bracket $\langle\cdot\rangle$ enclosing a single quantity denotes the expectation or mean value.

If the state vector $|\mathbf{x}\rangle$ is normalized clearly (8.4) becomes

$$\langle F\rangle = \langle\mathbf{x}|\mathbf{F}|\mathbf{x}\rangle \tag{8.5}$$

Substituting (8.1) into (8.5) yields for the expectation value

$$\begin{aligned}\langle F\rangle &= \sum_i (c_i^* c_i)F_{ii} \\ &= \sum_i (c_i^* c_i)\lambda_i \end{aligned}\tag{8.6}$$

where the λ_i are the eigenvalues of $\mathbf{F}$. Interpreting the coefficients c_i of each λ_i as a "weighting" factor or the probability of measuring the eigenvalue λ_i we can write

$$\langle F\rangle = \sum_i P_i\lambda_i \tag{8.7}$$

where

$$P_i = c_i^* c_i = |c_i|^2$$

In (8.7) we require the normalization

$$\sum_i |c_i|^2 = 1 \tag{8.7a}$$

for

$$c_i \geqslant 0 \tag{8.7b}$$

From (8.3) it is seen that the probability P_i is the square of the absolute value of the scalar product of the state vector $|\mathbf{x}\rangle$ and the ith eigenvector; i.e.,

$$P_i = |c_i|^2 = |\langle \mathbf{x}_i | \mathbf{x} \rangle|^2$$

If the system is in a state represented by one of the eigenvectors of the observable $\mathbf{F}$ then

$$P_i = |\langle \mathbf{x}_i | \mathbf{x}_j \rangle|^2 = \delta_{ij} \tag{8.8}$$

Hence the measured value of an observable $\mathbf{F}$ possesses a well defined probability of one or zero if the state of the system is an eigenvector of $\mathbf{F}$. In this case all probabilities are zero except the one measuring the occurrence of the eigenvalue λ_i, which is unity.

In considering two measurable quantities, say F and G, it can easily be shown that they can be measured simultaneously only if their associated resoec-tive operators $\mathbf{F}$ and $\mathbf{G}$ commute; i.e.,

$$[\mathbf{FG} - \mathbf{GF}] = 0$$

$$[\mathbf{F},\mathbf{G}] = 0$$

This is to say that the vector representing the system state must be an eigenvector of both observables. The quantities corresponding to the two operators can therefore be aribtrarily well defined in the same state. If the operators corresponding to the two physical quantities F and G do not commute, there will be a dispersion ΔF and ΔG inherent in their measurement. This dispersion is given by the inequality specified by the Heisenberg uncertainty principle:

$$\langle (\Delta F)^2 \rangle \langle (\Delta G)^2 \rangle \geqslant \frac{1}{4} \langle C \rangle^2$$

where

$$\mathbf{C} \equiv j[\mathbf{F},\mathbf{G}]$$

POSTULATE D. *The dynamical behavior of the system represented by the state vector* $|\mathbf{x}\rangle$ *is determined by the Schrodinger equation*

$$j\hbar\frac{\partial}{\partial t}|\mathbf{x}\rangle = \mathbf{H}|\mathbf{x}\rangle \qquad (8.9)$$

where $\hbar = h/2\pi = 1.05 \times 10^{-27}$ *erg sec and* $\mathbf{H}$ *is the operator (observable) corresponding to the classical Hamiltonian of the system.*

The relationship between measurable quantities and classical concepts can be readily established from the time development of the expectation value $\langle F\rangle$. In view of (8.9) we particularly want to explore the time dependency of $j\hbar\langle F\rangle$. From equation (8.5)

$$j\hbar\langle F\rangle = j\hbar\langle\mathbf{x}|\mathbf{F}|\mathbf{x}\rangle$$

Using the product rule for differentiation and the properties of the complex scalar product we have

$$j\hbar\frac{d}{dt}\langle\mathbf{x}|\mathbf{F}|\mathbf{x}\rangle = \langle\mathbf{x}|\mathbf{F}\, j\hbar\frac{\partial}{\partial t}|\mathbf{x}\rangle + j\hbar\langle\mathbf{x}|\frac{\partial\mathbf{F}}{\partial t}|\mathbf{x}\rangle - j\hbar\frac{\partial}{\partial t}\langle\mathbf{x}|\mathbf{F}|\mathbf{x}\rangle$$

which becomes upon using the operator relationship of (8.9)

$$j\hbar\frac{d}{dt}\langle\mathbf{x}|\mathbf{F}|\mathbf{x}\rangle = \langle\mathbf{x}|\mathbf{F}|\mathbf{x}\rangle - \langle\mathbf{x}|\mathbf{HF}|\mathbf{x}\rangle + j\hbar\langle\mathbf{x}|\frac{\partial\mathbf{F}}{\partial t}|\mathbf{x}\rangle$$

$$= \langle\mathbf{x}|\mathbf{FH} - \mathbf{HF}|\mathbf{x}\rangle + j\hbar\langle\mathbf{x}|\frac{\partial\mathbf{F}}{\partial t}|\mathbf{x}\rangle \qquad (8.10)$$

If the operator $d\mathbf{F}/dt$ is defined as

$$\frac{d\mathbf{F}}{dt} \equiv \langle\mathbf{x}|\frac{d\mathbf{F}}{dt}|\mathbf{x}\rangle$$

then equation (8.10) yields the dynamical law in operator form:

$$\frac{d\mathbf{F}}{dt} = \frac{1}{j\hbar}[\mathbf{F},\mathbf{G}] + \frac{\partial\mathbf{F}}{\partial t} \qquad (8.11)$$

Equation (8.11) is also referred to as the Heisenberg equation of motion.

8.3 ENERGY PICTURE

Clearly from (8.10) and (8.11) if the operators $\mathbf{F}$ and $\mathbf{H}$ commute, and $\mathbf{F}$ does not change with time, then the mean value of the measurable quantity F remains constant; i.e., $d\langle F\rangle/dt = 0$. It follows that, for this situation, the probability $|c_i|^2$ of measuring eigenvalue λ_i is also independent of time. Hence the observable is a conserved quantity. Of particular interest among conserved quantities is the associated energy. If the observable $\mathbf{H}$ corresponding to the classical Hamiltonian of the system is not a function of time ($d\mathbf{H}/dt = 0$), and since $\mathbf{H}$ commutes with itself, the mean or average value of the energy is constant; i.e.,

$$\frac{d\mathbf{H}}{dt} = \frac{1}{j\hbar}[\mathbf{H},\mathbf{H}] + \frac{\partial \mathbf{H}}{\partial t} = 0$$

Thus for an isolated system where $\partial\mathbf{H}/\partial t = 0$ equation (8.11) is in concert with the law of conservation of energy.

For conservative systems general solutions to the Schrodinger equation can be obtained in a straight forward manner. These solutions express the state vector in terms of the (energy) eigenvectors of $\mathbf{H}$. This can be seen by considering the following. From Postulate D the equation of motion is concisely

$$j\hbar\frac{\partial}{\partial t}|\mathbf{x}\rangle = \mathbf{H}|\mathbf{x}\rangle$$

Since t does not appear explicitly in the differential equation we can look for solutions of the form

$$|\mathbf{x}(t)\rangle = f(t)|\mathbf{x}\rangle$$

i.e., a form consisting of the product of a time-dependent part and a nontime-dependent part. The corresponding eigenvalue problem is formulated as

$$\begin{aligned}\mathbf{H}|\mathbf{x}_i\rangle &= \lambda_i|\mathbf{x}_i\rangle \\ &= E_i|\mathbf{x}_i\rangle \end{aligned} \tag{8.12}$$

where the observable $\mathbf{H}$, which is the measurable energy, is independent of time. The E_i represent the energy eigenvalues and $|\mathbf{x}_i\rangle$ the corresponding energy eigenvectors. Since the eigenvectors form a complete orthonormal set the vector $|\mathbf{x}(t)\rangle$ in the state space can be expanded as

$$|\mathbf{x}(t)\rangle = \sum_i a_i(t)|\mathbf{x}_i\rangle \tag{8.13}$$

where the coefficients $a_i(t)$ are time dependent. Substituting (8.13) into (8.9) gives

$$\sum_i \left(\mathbf{H} + \frac{\hbar}{j}\frac{\partial}{\partial t}\right) a_i(t)|\mathbf{x}_i\rangle = 0 \tag{8.14}$$

which, in view of (8.12), becomes

$$\sum_i \left[a_i(t)E_i + \frac{\hbar}{j}\frac{d}{dt}a_i(t)\right]|\mathbf{x}_i\rangle = 0 \tag{8.15}$$

The complex scalar product formed by (8.15) and the reciprocal vector $\langle \mathbf{x}_k|$ yields for the kth coefficient the first-order differential equation

$$a_k(t)E_k + \frac{\hbar}{j}\frac{d}{dt}a_k(t) = 0$$

for which the solution is

$$a_k(t) = c_k e^{-jE_k t/\hbar} \tag{8.16}$$

The coefficient c_k is a constant. Thus for an isolated system where the Hamiltonian is independent of time the solution to the dynamical equation (8.9) is of the form

$$|\mathbf{x}(t)\rangle = \sum_i c_i e^{-jE_i t/\hbar}|\mathbf{x}_i\rangle \tag{8.17}$$

Time enters into the solution of (8.9) strictly as a phase factor. The coefficients c_i are constants, the $|\mathbf{x}_i\rangle$ are eigenvectors of the energy operator, and the E_i are the corresponding energy eigenvalues. From (8.7) and (8.16) the probability of measuring the kth energy eigenvalue E_k is

$$P_k = a_k^* a_k = |c_k|^2 = \text{const.} \tag{8.18}$$

Hence for isolated systems all probabilities are constant in time. Only the relative phases of the component states change. If the system is in a given energy eigenstate at time t it will remain in that state corresponding to the same energy eigenvalue for all time.[1] This is true for the values of any other observables

1. *Op. cit.*

which commute with **H**. Thus states represented by solutions of the form of (8.17) are said to be *stationary states*. However, the name is somewhat misleading. It is not the state (8.17) that is independent of time, but the probability amplitude (8.18).

We also see from (8.17) that a stationary state has a well defined energy. E_i is a definite energy value in addition to being the expectation value. A determination of the energy of a system in a stationary state yields a particular value of E and only that value. The time-energy uncertainty relation of Postulate C, where

$$\Delta E \Delta t > \hbar$$

implies that a state with a precise energy ($\Delta E = 0$) is possible only if one has unlimited time to measure same. This is characteristic of stationary states in view of the constancy of the probability amplitude.

8.4 TIME-DEPENDENT PERTURBATIONS; STATE TRANSITIONS

In the discussions thus far we have examined mathematical problems that can be solved exactly. In practice problems that fit the theory and can be solved exactly are rare. One must therefore resort to methods of approximation. A powerful method of approximation is found in *perturbation* theory. The Hamiltonian is formed in two parts: one is large and characterizes the system for which the Schrodinger equation can be solved exactly, the other is small and acts as a *perturbation*. There are many physical problems of this kind. For example, systems which exert weak forces on one another, or time-varying external forces acting on a system. The system reactions can be described in terms of the unperturbed states of noninteracting systems for which exact solutions can be found. This technique suggests *transitions* among states. Thus perturbation theory provides a connection between the observables of a system and its stationary states.

In the previous section we discussed isolated systems for which the Hamiltonian was time invariant. The system state as a function of time was shown to be

$$|\mathbf{x}(t)\rangle = \sum_i c_i e^{-jE_i t/\hbar} |\mathbf{x}_i\rangle$$

The amplitude of the kth state is

$$a_k(t) = \langle \mathbf{x}_k | \mathbf{x}(t)\rangle = c_k e^{-jE_k t/\hbar}$$

from which it follows that

$$a_k^* a_k = |c_k|^2 = \text{const.}$$

If a time varying force acts upon the system the situation is changed. The Hamiltonian now contains a time dependent as well as a time invariant part. The coefficients a_i depend upon time in amplitude as well as phase. Consequently, it is possible that certain states will grow and/or decay with time. A system can change its character under the influence of an external force; the force produces transitions from $|\mathbf{x}_i\rangle$ to $|\mathbf{x}_i'\rangle$. This implies that the system energy changes from E_i to E_i'; the difference being the work done on the system by the external force.

Describing an external force as an additional time-dependent term in the Hamiltonian is, at best, only an approximation. For purposes of this discussion we will consider only the first-order term of perturbation theory. Also, we will assume that the time-dependent perturbation is weak and comes about as a result of the system *interacting with another system*. Thus the Hamiltonian can be approximated as

$$\mathbf{H} = \mathbf{H}^0 + \mathbf{V}(t) \tag{8.19}$$

where $\mathbf{H}^0$ is the Hamiltonian of the unperturbed system and $\mathbf{V}(t)$ is the small interaction or perturbation term. The superscript will be used to identify quantities associated with the unperturbed system. Further, we assume the eigenvectors $|\mathbf{x}_i^0\rangle$ of $\mathbf{H}^0$ are known and satisfy the relation

$$\mathbf{H}^0|\mathbf{x}_i^0\rangle = E_i^0|\mathbf{x}_i^0\rangle \tag{8.20}$$

where the corresponding energy eigenvalues E_i^0 are discrete. If the initial state at time t_0 is expanded in terms of the unperturbed eigenstates we have from (8.17)

$$|\mathbf{x}(t_0)\rangle = \sum_i c_i e^{-jE_i t_0/\hbar}|\mathbf{x}_i^0\rangle$$

In the absence of any disturbance, we have for all time

$$|\mathbf{x}(t)\rangle = \sum_i c_i e^{-jE_i^0 t/\hbar}|\mathbf{x}_i^0\rangle$$

However, in the presence of $\mathbf{V}(t)$ (8.17) is no longer the correct solution to the Schrodinger equation. To obtain the proper solution we must expand $|\mathbf{x}(t)\rangle$ at every instant of time with the amplitude coefficients c_i depending on time:

$$|\mathbf{x}(t)\rangle = \sum_i c_i(t) e^{-jE_i^0 t/\hbar} |\mathbf{x}_i^0\rangle \tag{8.21}$$

The probability amplitude of finding the system in the kth unperturbed state is

$$c_k(t) = \langle \mathbf{x}_k^0 | \mathbf{x}(t)\rangle e^{jE_k^0 t/\hbar} \tag{8.22}$$

To readily determine the manner in which state transitions occur we examine the time dependency of c_k. Substituting (8.21) into the equation of motion

$$j\hbar \frac{\partial}{\partial t} |\mathbf{x}\rangle = \mathbf{H}|\mathbf{x}\rangle = (\mathbf{H}^0 + \mathbf{V})|\mathbf{x}\rangle$$

gives

$$j\hbar \sum_i \left(\frac{d}{dt} c_i(t) - \frac{j}{\hbar} E_i^0 c_i(t) \right) |\mathbf{x}_i^0\rangle e^{-jE_i^0 t/\hbar} = \sum_i c_i(t) \mathbf{H} |\mathbf{x}_i^0\rangle e^{-jE_i^0 t/\hbar}$$

Forming the scalar product of the above with $\langle \mathbf{x}_k^0|$ results in

$$\left(j\hbar \frac{dc_k(t)}{dt} + E_k^0 c_k(t) \right) e^{-jE_k^0 t/\hbar} = \sum_i c_i(t) \langle \mathbf{x}_k^0 | \mathbf{H}^0 | \mathbf{x}_i^0 \rangle e^{-jE_i^0 t/\hbar}$$

$$+ \sum_i c_i(t) \langle \mathbf{x}_k^0 | \mathbf{V} | \mathbf{x}_i^0 \rangle e^{-jE_i^0 t/\hbar} \tag{8.23}$$

Since the $|\mathbf{x}_i^0\rangle$ form an orthonormal set Eq. (8.20) gives

$$\langle \mathbf{x}_k^0 | \mathbf{H} | \mathbf{x}_i^0 \rangle = E_i^0 \langle \mathbf{x}_k^0 | \mathbf{x}_i^0 \rangle = E_i^0 \delta_{ki}$$

which reduces (8.23) to

$$j\hbar \frac{d}{dt} c_k(t) = \sum_i c_i(t) \langle \mathbf{x}_k^0 | \mathbf{V} | \mathbf{x}_i^0 \rangle e^{-j(E_i^0 - E_k^0)t/\hbar}$$

$$= \sum_i c_i(t) V_{ki} e^{-j(E_i^0 - E_k^0)t/\hbar} \tag{8.24}$$

where

$$V_{ki} = \langle \mathbf{x}_k^0 | \mathbf{V} | \mathbf{x}_i^0 \rangle \tag{8.25}$$

Equation (8.24) is a system of simultaneous linear homogeneous differential equations. It expresses the equation of motion in terms of the eigenvectors of the unperturbed Hamiltonian $\mathbf{H}^0$. In matrix form (8.24) can be written as:

$$j\hbar \frac{d}{dt}\begin{bmatrix} c_1 \\ c_2 \\ \vdots \end{bmatrix} = \begin{bmatrix} V_{11} & V_{12}e^{-j(E_2^0 - E_1^0)t/\hbar} & \cdots \\ V_{21}e^{-j(E_1^0 - E_2^0)t/\hbar} & V_{22} & \cdots \\ \vdots & \vdots & \vdots \end{bmatrix}\begin{bmatrix} c_1 \\ c_2 \\ \vdots \end{bmatrix}$$

As yet no approximations have been made in arriving at (8.24). It is in the solution of this complicated set of equations that approximations are invoked. The solutions to (8.24) depend highly on the initial conditions. For simplicity it is assumed that at the initial time $t_0 = -\infty$ the system is definitely in one of the stationary states of the unperturbed Hamiltonian, say the ith state. We want to examine the probability of the system transitioning to state k. We assume that $\mathbf{H}^0$ has discrete energy levels. Thus the initial conditions for the time-dependent probability amplitudes become

$$c_i(-\infty) = 1 \qquad c_k(-\infty) = 0 \tag{8.26}$$

We begin our solution to (8.24) through successive approximations. Substituting in (8.24) the initial values of the coefficients c_i we have at time $t_0 = -\infty$

$$j\hbar \frac{d}{dt} c_k(t) = V_{ki} e^{-j(E_i^0 - E_k^0)t/\hbar} \tag{8.27}$$

Equation (8.27) is only valid for t such that

$$c_k(t) \ll c_i(t) \sim 1 \qquad (k \neq i)$$

We again make use of the initial conditions (8.26) to integrate (8.27):

$$c_k(t) = -\frac{j}{\hbar}\int_{-\infty}^{t} V_{ki} e^{-j(E_i^0 - E_k^0)t/\hbar}\, dt \qquad (k \neq i) \tag{8.28}$$

For small perturbations the $c_k(t)$ may remain small throughout. After the perturbations have stopped the system settles down to constant values of c_k evaluated at $t = +\infty$:

$$c_k(t) = -\frac{j}{\hbar}\int_{-\infty}^{\infty} V_{ki}\, e^{-j(E_i^0 - E_k^0)t/\hbar}\, dt \tag{8.29}$$

From formula (8.29) it is seen that a system influenced by a time-dependent perturbation makes transitions to other energy eigenstates of $\mathbf{H}^0$. The quantity $|c_k(\infty)|^2$ defines the transition probability from state i to k. It is proportional to the square of the absolute value of the Fourier component of the perturbation matrix element V_{ki}, evaluated at the *transition frequency* ω_{ki}; i.e.,

$$|c_k(\infty)|^2 = \left| -\frac{j}{\hbar}\int_{-\infty}^{\infty} V_{ki}\, e^{-j\omega_{ki}t}\, dt \right|^2$$

where ω_{ki} is deduced from the relation

$$\omega_{ki} = 2\pi f = \frac{E_i^0 - E_k^0}{\hbar}$$

If the system is initially in the higher energy state k the transition probability to the lower state i is, from (8.29),

$$c_i(\infty) = -\frac{j}{\hbar}\int_{-\infty}^{\infty} V_{ik}\, e^{-j\omega_{ik}t}\, dt$$

It can be shown that $\mathbf{V}$ is a Hermitian perturbation operator, and that $\omega_{ki} = -\omega_{ik}$. Thus it readily follows that

$$c_i(\infty) = -c_k^*(\infty) \tag{8.30}$$

The two transition probabilities are equal. Property (8.30) is the condition for *detailed balancing*. The energy difference $E = \hbar\omega_{ki}$ is transferred to the radiation field of the system.

8.5 CONSTANT PERTURBATION

As a matter of practical interest we consider the case where **V** is constant or varies slowly over the period $1/\omega_{ki}$. The system is in an initial unperturbed eigenstate $|\mathbf{x}_i^0\rangle$ at time $t_0 = 0$ and then subjected to a weak perturbation which persists at some near constant value. In keeping with our previous discussion the time development of a system with Hamiltonian $\mathbf{H} = \mathbf{H}^0 + \mathbf{V}$ can be conveniently described in terms of transitions between eigenstates of the unperturbed Hamiltonian $\mathbf{H}^0$. The approximate equations (8.27) apply:

$$j\hbar \frac{d}{dt} c_k(t) = V_{ki} e^{-j\omega_{ki} t}$$

Treating **V** as a constant and specifying the initial and final discrete states as i and k, respectively, integration of (8.27) gives

$$c_k(t) = \frac{\langle \mathbf{x}_k^0 | \mathbf{V} | \mathbf{x}_i^0 \rangle}{E_k^0 - E_i^0} (1 - e^{j\omega_{ki} t})$$

$$= \frac{V_{ki}}{E_k^0 - E_i^0} (1 - e^{j\omega_{ki} t}) \tag{8.31}$$

where

$$c_k(0) = 0, \qquad c_i(0) = 1, \qquad k \neq i$$

The probability that the system, in the initial state i at time $t_0 = 0$, will be in the final unperturbed eigenstate k $(k \neq i)$ at time t is

$$|c_k(t)|^2 = \frac{4|V_{ki}|^2}{(E_k^0 - E_i^0)^2} \sin^2 \left(\frac{\omega_{ki} t}{2} \right) \tag{8.32}$$

Equation (8.32) is illustrated graphically in Figure 8.1. It is a periodic function of t with a period equal to π/ω_{ki} and a peak at $\omega_{ki} = 0$. The expression is valid only as long as $c_i(t)$ can be approximated as $c_i(t) \approx 1$, during which time the transition probability to states where $E_k^0 \neq E_i^0$ remains small for weak perturbations. The probability of finding the system in state k is small unless the energy of the kth state is close to the energy of the initial state. However, transitions to

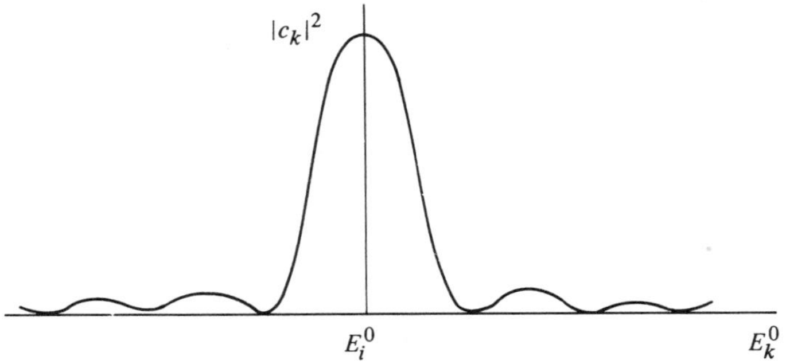

Figure 8.1

states where $E_k^0 \approx E_i^0$ have an important property. When $E_k^0 \approx E_i^0$ equation (8.32) can be approximated as

$$|c_k(t)|^2 \simeq \frac{1}{\hbar^2}|V_{ki}|^2 t^2 \tag{8.33}$$

Thus the transition probability to the kth unperturbed eigenstate increases quadratically with time. This has special importance when the states in the neighborhood of the initial energy are very closely spaced and constitute a near continuum.

It is not physically possible to measure $|c_k|^2$ for a single value of k. The classical measurement is of the *total* probability that the system made a transition from an initial to a final state. We define the total transition probability to all possible final states as

$$\text{Transition probability} \equiv \sum_k |c_k(t)|^2$$

where the summation extends over all final states under consideration. For a quasi-continuum of energy states per unit energy level we introduce the density of final unperturbed states denoted by $\rho(E)$. The quantity $\rho(E)dE$ measures the number of final states in the interval dE containing energy E. The total transition probability into these states is determined by multiplying (8.32) by $\rho(E_k^0)dE_k^0$ and integrating with respect to dE_k^0:

$$\int |c_k(t)|^2 \rho(E_k^0)\, dE_k^0 = 2\int |V_{ki}|^2 \frac{1 - \cos \omega_{ki} t}{(E_k^0 - E_i^0)^2} \rho(E_k^0)\, dE_k^0 \tag{8.34}$$

where

$$|V_{ki}| = |\langle \mathbf{x}_k^0 | \mathbf{V} | \mathbf{x}_i^0 \rangle|$$

The time rate of change of the total transition probability, w, is

$$w = \frac{d}{dt}\int |c_k(t)|^2 \rho(E_k^0)dE_k^0 = \frac{2}{\hbar}\int |V_{ki}| \frac{\sin \omega_{ki} t}{\omega_{ki}} \rho(E_k^0)dE_k^0 \qquad (8.35)$$

In analyzing (8.35) V_{ki} and ρ are reasonably constant over a small energy range δE_i^0 near E_i^0. However, $\sin \omega_{ki} t/\omega_{ki}$ oscillates rapidly in this same energy interval for all t satisfying the relation

$$t \gg \hbar/\delta E_i^0 \qquad (8.36)$$

and has a pronounced peak at $E_k^0 = E_i^0$. Clearly those transitions which tend to conserve the unperturbed energy are dominant. Also δE_i^0 is usually comparable to E_i^0. Thus $\hbar/\delta E_i^0$ is a very short time. Hence there is a large range of t where (8.36) is fulfilled yet the initial state i is not appreciably depleted. During this time (8.35) can be approximated as

$$w = \frac{d}{dt}\int_{-\infty}^{\infty} |c_k(t)|^2 \rho(E_k^0)dE_k^0 = \frac{2}{\hbar}|V_{ki}|\rho(E_k^0)\int_{-\infty}^{\infty} \frac{\sin \omega_{ki} t}{\omega_{ki}} d\omega_{ki} \quad (8.37)$$

Under the conditions stated the transition probability per unit time becomes

$$w = \frac{2\pi}{\hbar}|V_{ki}|\rho(E_k^0) \qquad (8.38)$$

Equation (8.38) shows a *constant* rate of transition. This result comes about because we summed over transitions which conserve the unperturbed energy ($E_k^0 = E_i^0$) and transitions that violate this conservation. From (8.33) it is seen that the transition probabilities of the former type increase quadratically with time, whereas from (8.32) it is seen that transitions of the latter type are periodic. The result is a compromise between these two and the transition rate is constant. Result (8.38) has been termed by Fermi as the *golden rule* of time-dependent perturbation theory.

8.6 EXPONENTIAL DECAY

Our discussions on perturbation theory was centered around the approximate solutions to equations (8.24):

$$j\hbar \frac{d}{dt} c_k(t) = \sum_i c_i(t) V_{ki} e^{j\omega_{ki} t}$$

From these equations it was seen that if $V_{ki} \neq 0$ transitions from an initial state to various available final states occur. The probability that the system will make a transition in time interval between t and $t + dt$ is equal to wdt. By the condition for detailed balancing, (8.30), these final states contribute to the probability amplitudes of the initial states through a feedback process. As the amplitudes of the final states grow, they do so at the expense of the initial states, since probability is conserved. Because of the different frequencies ω_{ki} of the feedback process the contributions made by the amplitudes c_k to amplitudes c_i are all of different phases. Thus if there are many available states k, forming a near-continuum, the contributions made by these states tend to cancel. This destructive interference in the probability amplitude is interpreted as a gradual (exponential) depletion of the initial state.

It is inferred above that the probability of finding the system at time t still in the initial state is proportional to $\exp(-wt)$. This is the exponential decay law. To derive the exponential decay law we no longer consider $c_i(t)$ on the right-hand side in (8.24) as a constant. However, in our approximation we will continue to neglect all other contributions to the change in $c_k(t)$. Equation (8.24) becomes

$$j\hbar \frac{d}{dt} c_k(t) = V_{ki}\, c_i(t')\, e^{j\omega_{ki}t'} \qquad k \neq i$$

$$c_k(t) = -\frac{j}{\hbar} V_{ki} \int_0^t c_i(t')\, e^{j\omega_{ki}t'}\, dt' \tag{8.39}$$

where $c_k(0) = 0$ for $k \neq i$. The prime identifies t with c_k. However, by reciprocal manipulation of (8.24) it can be shown that the equation of motion for $c_i(t)$ is rigorously

$$j\hbar \frac{d}{dt} c_i(t) = \sum_{k \neq i} V_{ki}\, c_k(t)\, e^{j\omega_{ik}t} + V_{ii}\, c_i(t) \tag{8.40}$$

Substituting (8.39) into (8.40) gives

$$\frac{d}{dt} c_i(t) = -\frac{1}{\hbar^2} \sum_{k \neq i} |V_{ki}|^2 \int_0^t c_i(t')\, e^{j\omega_{ki}(t'-t)}\, dt' - \frac{j}{\hbar} V_{ii}\, c_i(t) \tag{8.41}$$

Assuming that the pertinent final states are in a near-continuum with a density of states $\rho(E_k^0)$ the sum can be replaced by an integral extending over all possible transition frequencies. Equation (8.41) becomes

$$\frac{d}{dt}c_i(t) = \frac{1}{\hbar^2}\int_{-\infty}^{\infty}|V_{ki}|^2\rho(E_k^0)d\omega_{ki}\int_0^t c_i(t')e^{j\omega_{ki}(t'-t)}dt' - \frac{j}{\hbar}V_{ii}c_i(t) \quad (8.42)$$

Equation (8.42) is a differential equation of the Volterra type. It can be approximated by the simple equation[1]

$$\frac{d}{dt}c_i(t) = \left(-\frac{w}{2} - \frac{j}{\hbar}\delta E_i^0\right)c_i(t) \quad (8.43)$$

where δE_i^0 is the shift of the unperturbed energy level E_i^0 due to *second-order* perturbation.[2] With $c_i(0) = 1$ equation (8.43) yields

$$c_i(t) = \exp\left(-\frac{w}{2} - \frac{j}{\hbar}\delta E_i^0\right) \quad (8.44)$$

which describes the *exponential decay*.

We next examine the probability that the system has decayed into state k. Substituting (8.44) into (8.39) and integrating:

$$c_k(t) = -\frac{j}{\hbar}V_{ki}\int_0^t \exp\left[-\frac{j}{\hbar}\; E_i^0 + \delta E_i^0 - E_k^0 - j\hbar\frac{w}{2}\; t'\right]dt'$$

$$= \frac{1 - \exp\left(-\hbar\frac{w}{2}t\right)\exp\left[-\frac{j}{\hbar}(E_i^0 + \delta E_i^0 - E_k^0)t\right]}{E_k^0 - (E_i^0 + \delta E_i^0) + j\hbar\frac{w}{2}}$$

The probability $|c_k|^2$ that the system has decayed into state k becomes

$$|c_k|^2 = |V_{ki}|^2\,\frac{1 - 2\exp\left(\frac{\Gamma}{2}t\right)\cos\left(\frac{E_i^0 + \delta E_i^0 - E_k^0}{\hbar}\right)t + \exp(-\Gamma t)}{(E_k^0 - E_i^0 - \delta E_i^0)^2 + \frac{\Gamma^2}{4}} \quad (8.45)$$

1. See V. F. Weisskopf and E. P. Wigner, Z. Physik, 63, 54 (1930).
2. The reader is reminded that our discussion on perturbation addressed only first-order terms. An exception is made in the derivation of equation (8.43).

where $\Gamma = \hbar w$. For t long compared to the state lifetime $1/\Gamma$ the transition probability is approximated as

$$|c_k|^2 = \frac{|V_{ki}|^2}{(E_k^0 - E_i^0 - \delta E_i^0)^2 + \dfrac{\Gamma^2}{4}} \tag{8.46}$$

Equation (8.46) is a bell-shaped curve. It has a pronounced peak at those final state energy levels E_k^0 equal to $E_i^0 + \delta E_i^0$. The width of the curve is equal to $\hbar w$.

Appendixes

Appendix A

DIRAC DELTA FUNCTIONS

The delta function $\delta(\xi - a)$ is a mathematically improper function having the following properties (in one dimension):

$$\delta(\xi - a) = 0 \qquad \text{for } \xi \neq a \tag{A.1}$$

$$\int_{-\infty}^{\infty} \delta(\xi - a)\, d\xi = 1 \qquad \text{if region of integration includes } \xi = a \tag{A.2}$$

$$\int_{-\infty}^{\infty} \delta(\xi - a) = 0 \qquad \text{if region of integration does not include } \xi = a \tag{A.3}$$

Higher order delta functions are defined as the derivative of $\delta(\xi - a)$:

$$\delta^n(\xi - a) = \frac{d^n\delta(\xi - a)}{dt^n} \tag{A.4}$$

The delta function of another function can be transformed according to the rule

$$\delta(f(\xi)) = \frac{1}{\frac{df}{d\xi}}\delta(\xi - \xi_0) \tag{A.5}$$

where $f(\xi_0) = 0$. For an aribtrary function $f(\xi)$ the sifting property of delta functions provides for the integral equations

$$\int_{-\infty}^{\infty} f(\xi)\delta(\xi - a)d\xi = f(a) \tag{A.6}$$

$$\int_{-\infty}^{\infty} f(\xi)\delta^n(\xi - a)d\xi = (-1)^n f^n(a) \tag{A.7}$$

It is difficult to physically imagine the delta function or the unit impulse as it is also referred-to. Qualitatively, it can be thought of as a small pulse of high magnitude and infinitesimally small duration (see Fig. A-1). We require that as the peak of the curve gets higher, the width gets narrower in such a way that the area under the curve remains constant (unity). Thus the unit impulse $\delta(t)$ can be regarded the limit as $\Delta\xi \to 0$ of the pulse $p(t)$ having width $\Delta\xi$ and height $1/\Delta\xi$.

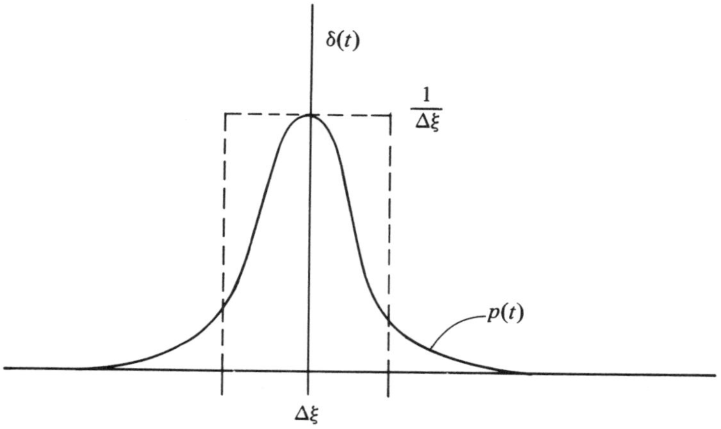

Figure A-1 Unit Impulse.

Appendix B

RESOLUTION OF CONTINUOUS-TIME SIGNALS INTO UNIT IMPULSES

An arbitrary signal $u(t)$ can be approximated in any finite time interval $-T \leqslant t \leqslant T$ by a finite number of unit pulses of width $\Delta\xi$ occurring at times $t = k\xi$, where $k = \pm 1, \pm 2, \ldots, \pm N = T/\Delta\xi$. Fig. B-1 illustrates the idea. Since the height of the unit pulse is $1/\Delta\xi$, the pulse at $t = k\Delta\xi$ is multiplied by $u(k\Delta\xi)\Delta\xi$ thereby resulting in the amplitude $u(t)$. The approximation of $u(t)$ can be written as

$$u(t) = \sum_{k=-N}^{N} u(k\Delta\xi)\Delta\xi\, p(t - k\Delta\xi) \tag{B.1}$$

$$= \lim_{\substack{\Delta\xi\to 0\\ N\to\infty}} \sum u(k\Delta\xi)\Delta\xi\, p(t - k\Delta\xi) \tag{B.2}$$

where $p(t)$ is defined in Appendix A. In the limit as $\Delta\xi \to 0$ and $N \to \infty$ the pulses become impulses and the summation becomes an integral. Thus the approximation becomes exact. We have

$$u(t) = \int_{-T}^{T} u(\xi)\delta(t - \xi)d\xi \tag{B.3}$$

over the finite time interval $(-T,T)$. Extending the integral over the entire time domain defining $u(t)$, i.e., letting $T \to \infty$,

$$u(t) = \int_{-\infty}^{\infty} u(\xi)\delta(t - \xi)d\xi \tag{B.4}$$

Thus, any continuous-time signal $u(t)$ can be resolved into a continuum of unit pulses. It follows that the response of a linear system to excitation $u(t)$ can be readily found if the response to the unit impulse is known. Hence the unit impulse response completely characterizes the system.

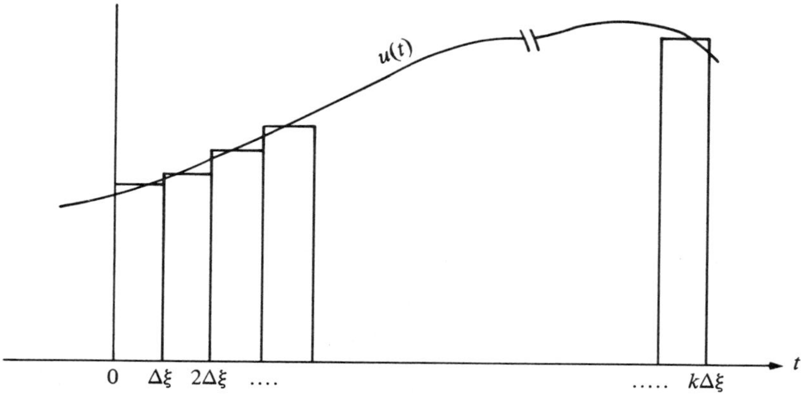

Figure B-1 Approximation of $u(t)$ by unit impulses.

Appendix C

DISCRETE-TIME STATE EQUATIONS

From (2.1) the standard form of the nth order difference equation can be written as

$$a_0 y(k) + a_1 y(k-1) + \ldots + a_n y(k-n) = b_1 u(k-1) + \ldots + b_s u(k-s) \quad \text{(C.1)}$$

$$(a_0 + a_1 A^{-1} + \ldots + a_n A^{-n}) y(k) = (b_1 A^{-1} + \ldots + b_s A^{-s}) u(k) \quad \text{(C.2)}$$

where A is the linear advance operator defined for any interger ν as

$$A^{\nu} f(k) = f(k+\nu) \quad \text{(C.3)}$$

Introducing the variable $v(k)$ where

$$(a_0 + a_1 A^{-1} + \ldots a_n A^{-n}) v(k) = u(k) \quad \text{(C.4)}$$

and

$$(b_1 A^{-1} + b_2 A^{-2} + \ldots + b_s A^{-s}) v(k) = y(k) \quad \text{(C.5)}$$

equation (C.2) gives

$$v(k) = \frac{1}{a_0}[u(k) - a_n v(k-n) - a_{n-1}v(k-n+1) - \dots - a_1 v(k-1)] \tag{C.6}$$

Specifying the elements of $\mathbf{x}(k)$ as

$$\begin{aligned} x_1(k) &= v(k-n) \\ x_2(k) &= v(k-n+1) \\ &\dots\dots\dots\dots \\ x_n(k) &= v(k-1) \end{aligned} \tag{C.7}$$

we can write

$$\begin{aligned} x_1(k+1) &= x_2(k) \\ x_2(k+1) &= x_3(k) \\ &\dots\dots\dots\dots \\ x_{n-1}(k+1) &= x_n(k) \\ x_n(k+1) &= \frac{1}{a_0}[u(k) - a_n x_1(k) - \dots - a_1 x_n(k)] \end{aligned} \tag{C.8}$$

In matrix form (C.8) becomes

$$\mathbf{x}(k+1) = \mathbf{A}\mathbf{x}(k) + \mathbf{B}u(k) \tag{C.9}$$

where

$$\mathbf{A} = \begin{bmatrix} 0 & 1 & 0 & \dots & 0 \\ 0 & 0 & 1 & \dots & 0 \\ \dots & \dots & \dots & \dots & \dots \\ -\frac{a_n}{a_0} & -\frac{a_{n-1}}{a_0} & & \dots & -\frac{a_1}{a_0} \end{bmatrix} \qquad \mathbf{B} = \begin{bmatrix} 0 \\ 0 \\ \dots \\ \frac{1}{a_0} \end{bmatrix} \tag{C.10}$$

From (C.5) we have

$$y(k) = b_1 v(k-1) + b_2 v(k-2) + \dots + b_s v(k-s) \tag{C.11}$$

Substituting (C.7) into (C.11) gives

$$y(k) = b_1 x_n(k) + b_2 x_{n-1}(k) + \dots + b_s x_{n+1-s}(k) \tag{C.12}$$

or, in matrix form,

$$\mathbf{y}(k) = \mathbf{C}\mathbf{x}(k) + \mathbf{D}\mathbf{u}(k) \tag{C.13}$$

where

$$\mathbf{C} = [b_n \ \dots \ b_1] \qquad \mathbf{D} = 0 \tag{C.14}$$

Matrices (C.10) and (C.14) are not unique. They are one of a variety of ways to represent (C.9) and (C.13).

Appendix D

TRANSFORMS

Listed below are $\mathcal{Z}$ transforms of some of the more commonly used mathematical functions:

$$\mathcal{Z}[f(k)] = F(z) = \sum_{k=0}^{\infty} f(k) z^{-k} \qquad k \geqslant 0$$

$$\mathcal{Z}[a^k] = \frac{1}{1 - az^{-1}} \qquad |z| > |a|$$

$$\mathcal{Z}[e^{j\omega k}] = \frac{1}{1 - e^{j\omega} z^{-1}} \qquad |z| > 1$$

$$\mathcal{Z}[\cos \omega k] = \frac{1 - z^{-1} \cos \omega}{1 - 2z^{-1} \cos \omega + z^{-2}} \qquad |z| > 1$$

$$\mathcal{Z}[\sin \omega k] = \frac{z^{-1} \sin \omega}{1 - 2z^{-1} \cos \omega + z^{-2}} \qquad |z| > 1$$

$$\mathcal{Z}[k^\nu f(k)] = \left(-z\frac{d}{dz}\right)^\nu F(z)$$

$$\mathcal{Z}[k] = \frac{z^{-1}}{(1-z^{-1})^2} \qquad |z| > 1$$

$$\mathcal{Z}[k^2] = \frac{z^{-1}(1-z^{-1})}{(1-z^{-1})^3} \qquad |z| > 1$$

$$\mathcal{Z}[k^{-1}] = -\log(1-z^{-1}) \qquad k \geqslant 1, \ |z| > 1$$

$$\mathcal{Z}\left[\frac{1}{k!}\right] = e^{1/z} \qquad |z| > 0$$

$$\mathcal{Z}[f(k+1)] = z[F(z) - f(0)]$$

$$\mathcal{Z}[f(k-m)] = z^{-m}F(z)$$

$$\mathcal{Z}[a^k f(k)] = F(a^{-1}z)$$

$\mathcal{Z}$ transforms involving difference operators can be written according to the rule

$$\mathcal{Z}[\Delta f(k)] = (z-1)F(z) - zf(0)$$

$$\mathcal{Z}[-\Delta f(k)] = (1-z^{-1})F(z)$$

where $\Delta f(k)$ is the forward difference operator defined as

$$\Delta f(k) = f(k+1) - f(k)$$

and $-\Delta f(k)$ is the backward difference operator defined as

$$-\Delta f(k) = f(k) - f(k-1)$$

The inversion of the $\mathcal{Z}$ transform can be accomplished in a variety of ways. Listed below are a few elementary transform terms useful in expansion by partial fractions:

$F(z)$	Time sequence: 1. $F(z)$ converges $\lvert z \rvert > a$	Time sequence: 2. $F(z)$ converges $\lvert z \rvert < a$
$\dfrac{1}{z-a}$	$a^{k-1}\big\rvert_{k \geqslant 1}$	$-a^{k-1}\big\rvert_{k \leqslant 0}$
$\dfrac{z}{(z-a)^2}$	$ka^{k-1}\big\rvert_{k \geqslant 1}$	$-ka^{k-1}\big\rvert_{k \leqslant 0}$
$\dfrac{z(z+a)}{(z-a)^3}$	$k^2a^{k-1}\big\rvert_{k \geqslant 1}$	$-k^2a^{k-1}\big\rvert_{k \leqslant 0}$
$\dfrac{z(z^2+4az+a^2)}{(z-a)^4}$	$k^3a^{k-1}\big\rvert_{k \geqslant 1}$	$-k^3a^{k-1}\big\rvert_{k \leqslant 0}$

Appendix E

ANALOGOUS QUANTITIES OF CONTINUOUS-TIME AND DISCRETE-TIME SYSTEMS

QUANTITY	CONT.-TIME SYSTEM	DISCRETE-TIME SYSTEM
Fundamental matrix	$\Phi(t) = e^{\mathbf{A}t}$	$\Phi(k) = \mathbf{A}^k$
Transform of fundamental matrix	$\Phi(s) = (s\mathbf{I} - \mathbf{A})^{-1}$	$\Phi(z) = (\mathbf{I} - \mathbf{A}z^{-1})^{-1}$
Impulse response matrix	$\mathbf{H}(t) = \mathbf{C}\Phi(t)\mathbf{B} + \mathbf{D}\delta(t)$	$\mathbf{H}(k) = \mathbf{C}\Phi(k-1)\mathbf{B} \quad (k \geqslant 1)$ $\mathbf{H}(k) = \mathbf{D} \quad (k = 0)$
Transfer function matrix	$\mathbf{H}(s) = \mathbf{C}\Phi(s)\mathbf{B} + \mathbf{D}$	$\mathbf{H}(z) = \mathbf{C}z^{-1}\Phi(z)\mathbf{B} + \mathbf{D}$

Appendix F

STOCHASTIC PROCESSES

Assume that a given experiment is performed on a system in which outcomes ζ are measured (see Figure F.1). Let the outcomes ζ form the space S. Associated with S are subsets called events and the probabilities of these events. To every outcome ζ we can assign the time function $\underline{u}(\zeta,t)$. The set of functions $\{\underline{u}(\zeta,t)\}$, one for each ζ, is called a *stochastic process*. For a specific outcome ζ_i the expression $\{\underline{u}(\zeta_i,t)\}$ is a single time function. For a specific time t_i, $\{\underline{u}(\zeta,t_i)\}$ depends on ζ and is a random variable. $\{\underline{u}(\zeta_i,t_i)\}$ is a number. Where it is understood that ζ is the random variable of the random process $\{\underline{u}(\zeta,t)\}$ we shall use the notation $\underline{u}(t)$ to represent the stochastic process:

$$\{u(\zeta,t)\} = \underline{u}(t) \tag{F.1}$$

The notation $\underline{u}$ is used to distinguish a sample function of a random process from the function u established in Chapter 1.

By repeating the experiment n times we obtain n sample functions as shown in Figure F.1. For a specific instant of time t_i we observe the values $\zeta \leqslant a$ and denote this total number of trials as $n_t(a)$. Accordingly, we establish the *distribution function* $F(a;t)$ as:

$$F(a;t) \simeq \frac{n_t(a)}{n}$$

$$= P\{\underline{u}(\zeta,t_i) \leqslant a\} \tag{F.2}$$

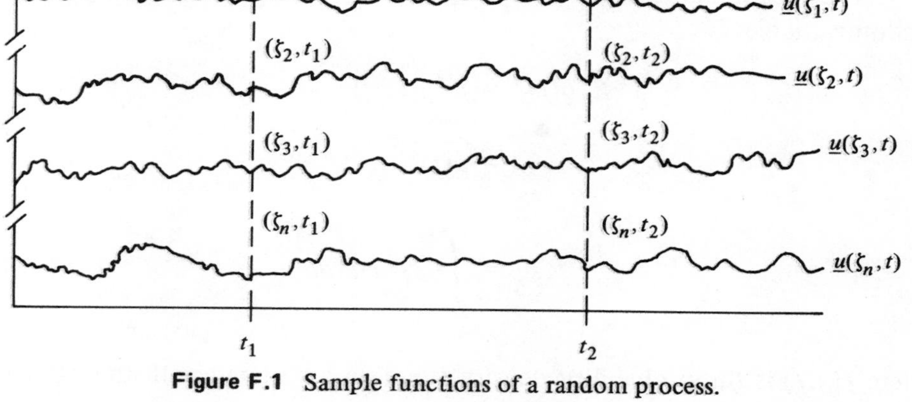

Figure F.1 Sample functions of a random process.

$F(a;t)$ is the probability of the event $\{\underline{u}(\zeta,t_i) \leqslant a\}$ consisting of all outcomes ζ such that at time t_i the functions $\underline{u}(t)$ do not exceed the value $\zeta = a$. Equation (F.2) expresses the *first order distribution* of the process $\underline{u}(t)$. The density of $F(a;t)$ is, simply,

$$\frac{\partial F(a;t)}{\partial a} = f(a;t) \tag{F.3}$$

Similarly, given two instants of time t_1 and t_2 where the corresponding values of ζ are a_1 and a_2, respectively, we form the *joint distribution function* $F(a_1,a_2;t_1,t_2)$ as follows:

$$F(a_1,a_2;t_1,t_2) = P\{\underline{u}(t_1) \leqslant a_1, \underline{u}(t_2) \leqslant a_2\} \tag{F.4}$$

Expression (F.4) is the *second order distribution* of the process $\underline{u}(t)$. The corresponding density is

$$\frac{\partial F(a_1,a_2;t_1,t_2)}{\partial a_1 \partial a_2} = f(a_1,a_2;t_1,t_2) \tag{F.5}$$

Although it is not entirely correct we will state that for "practical" purposes a real stochastic process can be statistically determined by its nth order distribution function

$$F(a_1,a_2,\ldots a_n;t_1,t_2,\ldots,t_n) = P\{\underline{u}(t_1) \leqslant a_1, \underline{u}(t_2) \leqslant a_2, \ldots \underline{u}(t_n) \leqslant a_n\}$$

The nth order density

$$f(a_1,a_2,\ldots,a_n;t_1,t_2,\ldots,t_n)$$

is determined by differentiating with respect to all variables a_i $(i = 1,\ldots,n)$.

The *mean value* of the process $\underline{u}(t)$ is the expected value $E\{\underline{u}(t)\}$ of the random variable:

$$\begin{aligned} E\{\underline{u}(t)\} &= E\{\underline{u}(\zeta,t)\} \\ &= \overline{\underline{u}(t)} \\ &= \int_{-\infty}^{\infty} a f(a;t)\,da \end{aligned} \tag{F.6}$$

where $f(a;t)$ is the probability density function associated with the value $\zeta = a$, and $f(a;t)da$ is the probability of finding the value a within $a \pm da$. Therefore, it must follow that:

$$\int_{-\infty}^{\infty} f(a;t)da = 1 \tag{F.7}$$

The *auto correlation* $R(t_1, t_2)$ of a process $\underline{u}(t)$ is the joint moment of $\{\underline{u}(\zeta, t_1)\}$ and $\{\underline{u}^*(\zeta, t_2)\}$, i.e., of $\underline{u}(t_1)$ and $\underline{u}^*(t_2)$:

$$\begin{aligned} R(t_1, t_2) &= E\{\underline{u}(t_1)\underline{u}^*(t_2)\} = \overline{\underline{u}(t_1)\underline{u}^*(t_2)} \\ &= R_u(t_1, t_2) \\ &= \int_{-\infty}^{\infty} a_1 a_2 f(a_1, a_2; t_1, t_2) da_1 da_2 \end{aligned} \tag{F.8}$$

where * denotes the conjugate. For real processes $\underline{u}(t) = u^*(t)$.

A stochastic process is said to be *stationary in the strict sense* if its statistics are independent of a shift in the time origin, i.e., $\underline{u}(t)$ and $\underline{u}(t + \delta)$ have the same statistics for all δ. Similarly, $\underline{u}(t)$ is said to be *stationary in the wide sense* if its expected value or mean is a constant and its autocorrelation function depends only on $t_2 - t_1$:

$$E\{\underline{u}(t)\} = \overline{\underline{u}(t)} = \text{const.} \tag{F.9}$$

$$\begin{aligned} R(t_1,t_2) = R(t_2 - t_1) = R(\xi) &= E\{\underline{u}(t - \xi)\underline{u}(t)\} \\ &= \overline{\underline{u}(t - \xi)\underline{u}(t)} \end{aligned} \tag{F.10}$$

where it is understood that t establishes a fixed time and ξ varies over the time interval of interest.

The process $\underline{u}(t)$ is said to have *uncorrelated*, *independent* or *orthogonal* increments if $\underline{u}(t_i)$ and $\underline{u}(t_{i+1})$ is a sequence of uncorrelated, independent or orthogonal intervals.

Given two real or complex processes $\underline{u}(t)$ and $\underline{v}(t)$ the *cross-correlation* of the two processes is defined as

$$\begin{aligned} R_{uv}(t_1,t_2) &= E\{\underline{u}(t_1)\underline{v}^*(t_2)\} \\ &= \overline{\underline{u}(t_1)\underline{v}^*(t_2)} \end{aligned} \tag{F.11}$$

The two processes are said to be *orthogonal* if

$$R_{uv}(t_1,t_2) = 0 \tag{F.12}$$

Similarly, the two processes are *uncorrelated* if their *cross-covariance* C_{uv} is zero, i.e., where

$$\begin{aligned} C_{uv}(t_1,t_2) &= R_{uv}(t_1,t_2) - E\{\underline{u}(t_1)\}E\{\underline{v}^*(t_2)\} \\ &= R_{uv}(t_1,t_2) - \overline{\underline{u}(t_1)}\,\overline{\underline{v}(t_2)} \\ &= 0 \end{aligned} \tag{F.13}$$

It follows that for two processes to be uncorrelated

$$R_{uv}(t_1,t_2) = \overline{\underline{u}(t_1)}\,\overline{\underline{v}(t_2)} \tag{F.14}$$

Lastly, the two processes are *independent* if $[\underline{u}(t_1),\ldots,\underline{u}(t_n)]$ is independent of $[\underline{v}(t'_1),\ldots,\underline{v}(t'_k)]$ for any $t_1,\ldots,t_n,t'_1,\ldots,t'_k$.

A real stationary stochastic process $\underline{u}(t)$ is said to be *ergodic* if all of its statistics can be determined from a single sample function $\underline{u}(\zeta_i, t)$. Since many statistical parameters can be expressed as time averages it can be said that $\underline{u}(t)$ is ergodic if the time averages are equal to the ensemble averages, i.e., if

$$\langle \underline{u}(\zeta_i,t)\rangle = \langle \underline{u}(t)\rangle = E\{\underline{u}(\zeta,t)\} = E\{\underline{u}(t)\} \tag{F.15}$$

where $\langle \underline{u}(\zeta_i,t)\rangle$ is the time average of the sample function $\underline{u}(\zeta_i, t)$. We imply in (F.15) that the statistics for $\underline{u}(t)$ are the same for all sample functions $\underline{u}(\zeta_i,t)$. Therefore,

$$\langle \underline{u}(\zeta,t)\rangle = \langle \underline{u}(t)\rangle$$

Below we examine some of the conditions under which (F.15) holds true.

Given a stationary stochastic process $\underline{u}(t)$ the limits

$$\langle \underline{u}(t)\rangle = \lim_{T\to\infty} \frac{1}{2T}\int_{-t}^{t} \underline{u}(t)\,dt \tag{F.16}$$

$$R(\xi) = \lim_{T\to\infty} \frac{1}{2T}\int_{-t}^{t} \underline{u}(t-\xi)\,\underline{u}(t)\,dt \tag{F.17}$$

where T is the interval $(-t, t)$, define the mean and autocorrelation as time averages. Form the ergodicity theorem every function $\underline{u}(\zeta, t_i)$ in (F.16) defines a number. Thus, $\langle \underline{u}(t) \rangle$ is a random variable. The random variable $\langle \underline{u}(t) \rangle$ is equal to the constant $E\{\underline{u}(t)\} = \overline{u(t)}$ only if its *variance* is zero. To determine this variance we form the *finite* average

$$\langle \underline{u}(t) \rangle = \frac{1}{2T} \int_{-t}^{t} \underline{u}(t)\, dt \tag{F.18}$$

which has the mean

$$E\{\langle \underline{u}(t) \rangle\} = \frac{1}{2T} \int_{-t}^{t} E\{\underline{u}(t)\} dt$$

$$= \frac{1}{2T} \int_{-\infty}^{\infty} \overline{\underline{u}(t)}\, dt \tag{F.19}$$

On forming the second moment $E\{|\langle \underline{u}(t) \rangle|^2\}$ it can be shown that the variance σ^2 of the time average $\langle \underline{u}(t) \rangle$ is

$$\sigma^2 = \frac{1}{T} \int_{0}^{2t} \left(1 - \frac{\xi}{2T}\right) [R(\xi) - |\overline{\underline{u}(t)}|^2] d\xi \tag{F.20}$$

$$= E\{|\langle \underline{u}(t) \rangle|^2\} - |\overline{\underline{u}(t)}|^2 \tag{F.21}$$

If (F.20) tends to zero as $T \to \infty$ then $\underline{u}(t)$ is ergodic with respect to its mean. Thus,

$$\langle \underline{u}(t) \rangle = E\{\underline{u}(t)\} = \overline{\underline{u}(t)}$$

Using (F.17) it can be shown, in a manner similar to that above, that the process $\underline{u}(t - \xi)\underline{u}(t)$ is ergodic with respect to its autocorrelation function

$$R(\xi) = \lim_{T \to \infty} \frac{1}{2T} \int_{-t}^{t} \underline{u}(t - \xi)\, \underline{u}(t)\, dt \tag{F.22}$$

$$= E\{\underline{u}(t - \xi)\underline{u}(t)\} = \overline{\underline{u}(t - \xi)\underline{u}(t)} \tag{F.23}$$

if the limit

$$\lim_{T\to\infty} \frac{1}{T}\int_0^{2T}\left(1 - \frac{\tau}{2T}\right)[R(\tau) - R^2(\xi)]\,d\xi = 0 \qquad \text{(F.24)}$$

where

$$R(\tau) = E\{\underline{u}(t+\xi+\tau)\underline{u}(t+\tau)\underline{u}(t+\xi)\underline{u}(t)\} \qquad \text{(F.25)}$$

The *power spectrum* or *spectral density* of a process $\underline{u}(t)$ is the Fourier transform $S(\omega)$ of its autocorrelation function $R(\xi)$:

$$S(\omega) = \mathcal{F}[R(\xi)]$$

$$= \int_{-\infty}^{\infty} R(\xi)e^{-j\omega\xi}\,d\xi \qquad \text{(F.26)}$$

From the inverse transform $R(\xi)$ can be expressed in terms of $S(\omega)$ as

$$R(\xi) = \frac{1}{2\pi}\int_{-\infty}^{\infty} S(\omega)e^{j\omega\xi}\,d\omega \qquad \text{(F.27)}$$

Setting $\xi = 0$ in (F.27) gives

$$R(0) = \frac{1}{2\pi}\int_{-\infty}^{\infty} S(\omega)d\omega = E\{|\underline{u}(t)|^2\} \geqslant 0 \qquad \text{(F.28)}$$

The power spectrum $S(\omega)$, as a time average (ergodicity) of the process $\underline{u}(t)$, is defined as the limit

$$S(\omega) = \lim_{T\to\infty} \frac{1}{2T}\left|\int_{-t}^{t} \underline{u}(t)e^{-j\omega\xi}\,d\xi\right|^2 \qquad \text{(F.29)}$$

of the random power

$$\underline{S}(\omega) = \frac{1}{2T}\left|\int_{-t}^{t} \underline{u}(t)e^{-j\omega\xi}\,d\xi\right|^2 \qquad \text{(F.30)}$$

For (F.29) to hold true we demand that $\underline{S}(\omega)$ tend to $S(\omega)$ and its variance tend to zero as $T \to \infty$.

Appendix G: TRANSFER FUNCTIONS & PLOTS OF TYPICAL NETWORKS & SYSTEMS

Transfer Functions of *RC* Input and Output Networks†

To generate a specific $f(s)$, rewrite $f(s)$ in the form Z_o/Z_i where Z_o and Z_i are each in a form corresponding to a function in the left-hand column. Choose input and output networks in accordance with the diagrams and relations adjacent to the function representing Z_i and Z_o, respectively.

Impedance function	Network	Relations	Inverse relations
A	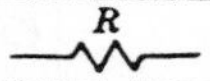	$A = R$	$R = A$
$\frac{A}{1+sT}$	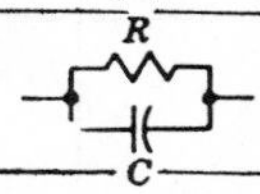	$A = R$ $T = RC$	$R = A$ $C = \frac{T}{A}$
$A(1+sT)$	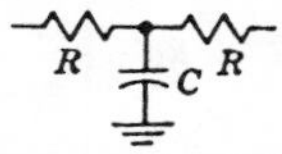	$A = 2R$ $T = \frac{RC}{2}$	$R = \frac{A}{2}$ $C = \frac{4T}{A}$
$A\left(\frac{1+s\theta T}{1+sT}\right)$ $\theta < 1$	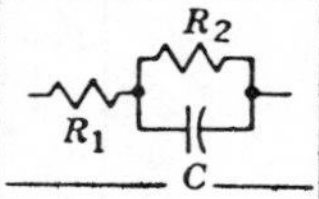	$A = R_1 + R_2$ $T = R_2C$ $\theta = \frac{R_1}{R_1+R_2}$	$R_1 = A\theta$ $R_2 = A(1-\theta)$ $C = \frac{T}{A(1-\theta)}$
	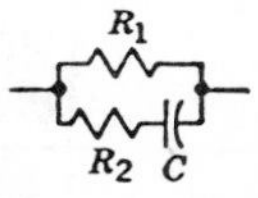	$A = R_1$ $T = (R_1 + R_2)C$ $\theta = \frac{R_2}{R_1+R_2}$	$R_1 = A$ $R_2 = \frac{A\theta}{1-\theta}$ $C = \frac{T(1-\theta)}{A}$
$A\left(\frac{1+sT}{1+s\theta T}\right)$ $\theta < 1$	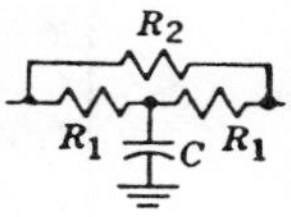	$A = \frac{2R_1R_2}{2R_1+R_2}$ $T = \frac{R_1C}{2}$ $\theta = \frac{2R_1}{2R_1+R_2}$	$R_1 = \frac{A}{2(1-\theta)}$ $R_2 = \frac{A}{\theta}$ $C = \frac{4T(1-\theta)}{A}$
	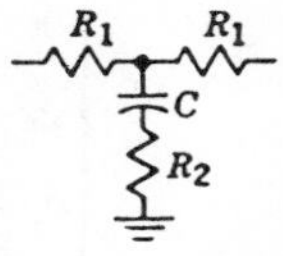	$A = 2R_1$ $T = \left(R_2 + \frac{R_1}{2}\right)C$ $\theta = \frac{2R_2}{2R_2+R_1}$	$R_1 = \frac{A}{2}$ $R_2 = \frac{A\theta}{4(1-\theta)}$ $C = \frac{4T(1-\theta)}{A}$
	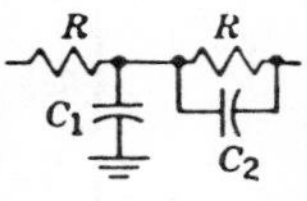	$A = 2R$ $T = \frac{R}{2}(C_1 + C_2)$ $\theta = \frac{2C_2}{C_1+C_2}$	$R = \frac{A}{2}$ $C_1 = \frac{2T(2-\theta)}{A}$ $C_2 = \frac{2T\theta}{A}$
$\frac{1}{sB}\left[\frac{(1+sT_1)(1+sT_3)}{1+sT_2}\right]$ $T_1 < T_2 < T_3$	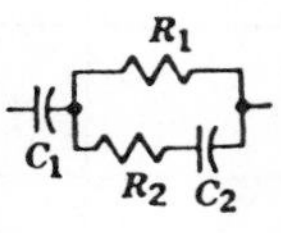	$B = C_1$ $T_2 = (R_1 + R_2)C_2$ $T_1T_3 = R_1R_2C_1C_2$ $T_1 + T_3 = R_1C_1 + R_2C_2 + R_1C_2$	$R_1 = \frac{T_1 + T_3 - T_2}{B}$ $R_2 = \frac{T_1T_3(T_1 + T_3 - T_2)}{B(T_3 - T_2)(T_2 - T_1)}$ $C_1 = B$ $C_2 = \frac{B(T_3 - T_2)(T_2 - T_1)}{(T_1 + T_3 - T_2)^2}$
	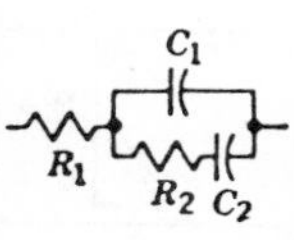	$B = C_1 + C_2$ $T_2 = R_2\frac{C_1C_2}{C_1+C_2}$ $T_1T_3 = R_1R_2C_1C_2$ $T_1 + T_3 = R_1C_1 + R_2C_2 + R_1C_2$	$R_1 = \frac{T_1T_3}{BT_2}$ $R_2 = \frac{(T_1T_2 + T_2T_3 - T_1T_3)^2}{BT_2(T_3 - T_2)(T_2 - T_1)}$ $C_1 = \frac{BT_2^2}{T_1T_2 + T_2T_3 - T_1T_3}$ $C_2 = \frac{B(T_3 - T_2)(T_2 - T_1)}{T_1T_2 + T_2T_3 - T_1T_3}$

Transfer Functions of RC Input and Output Networks (*Continued*)

Impedance function	Network	Relations	Inverse relations
$\frac{1}{sB}\left[\frac{(1+sT_1)(1+sT_3)}{1+sT_2}\right]$ $T_1 < T_2 < T_3$	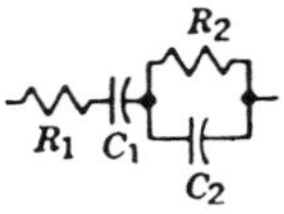	$B = C_1$ $T_2 = R_2C_2$ $T_1T_3 = R_1R_2C_1C_2$ $T_1 + T_3 = R_1C_1 + R_2C_2 + R_2C_1$	$R_1 = \frac{T_1T_3}{BT_2}$ $R_2 = \frac{(T_3 - T_2)(T_2 - T_1)}{BT_2}$ $C_1 = B$ $C_2 = \frac{BT_2^2}{(T_3 - T_2)(T_2 - T_1)}$
	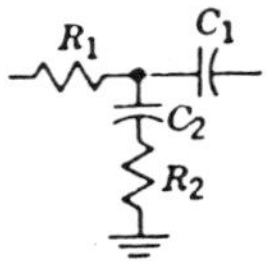	$B = C_1$ $T_2 = R_2C_2$ $T_1T_3 = R_1R_2C_1C_2$ $T_1 + T_3 = R_1C_1 + R_2C_2 + R_1C_2$	$R_1 = \frac{T_1T_3}{BT_2}$ $R_2 = \frac{T_1T_2T_3}{B(T_3 - T_2)(T_2 - T_1)}$ $C_1 = B$ $C_2 = \frac{B(T_3 - T_2)(T_2 - T_1)}{T_1T_3}$
$\frac{1}{sB}(1+sT_1)(1+sT_2)$ $T_1 \neq T_2$	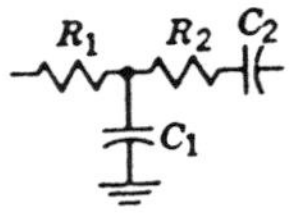	$B = C_2$ $T_1T_2 = R_1R_2C_1C_2$ $T_1 + T_2 = R_1C_1 + R_2C_2 + R_1C_2$	$R_1 = \frac{(\sqrt{T_1} - \sqrt{T_2})^2}{B}$ $R_2 = \frac{\sqrt{T_1T_2}}{B}$ $C_1 = \frac{B\sqrt{T_1T_2}}{(\sqrt{T_1} - \sqrt{T_2})^2}$ $C_2 = B$
$\frac{1}{sB}\left[\frac{(1+sT_1)(1+sT_2)}{s\sqrt{T_1T_2}}\right]$ $T_1 \neq T_2$	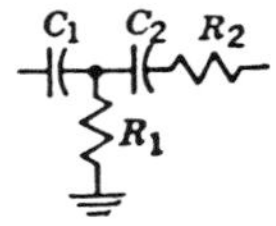	$B = C_2$ $T_1T_2 = R_1R_2C_1C_2$ $T_1 + T_2 = R_1C_1 + R_2C_2 + R_1C_2$	$R = \frac{(\sqrt{T_1} - \sqrt{T_2})^2}{B}$ $R_2 = \frac{\sqrt{T_1T_2}}{B}$ $C_1 = \frac{B\sqrt{T_1T_2}}{(\sqrt{T_1} - \sqrt{T_2})^2}$ $C_2 = B$
$\frac{1}{sB}\left[\frac{(1+sT_1)(1+sT_2)}{s^2T_1T_2}\right]$ $T_1 < T_2$	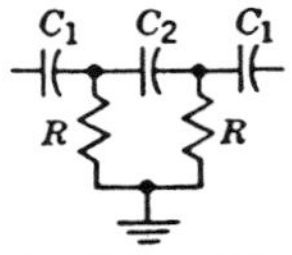	$B = \frac{C_1C_2}{C_1 + 2C_2}$ $T_1 = RC_1$ $T_2 = R(C_1 + 2C_2)$	$R = \frac{T_1(T_2 - T_1)}{2BT_2}$ $C_1 = \frac{2BT_2}{T_2 - T_1}$ $C_2 = \frac{BT_2}{T_1}$
$A\left(\frac{1+sT_1}{1+s^2T_1T_2}\right)$	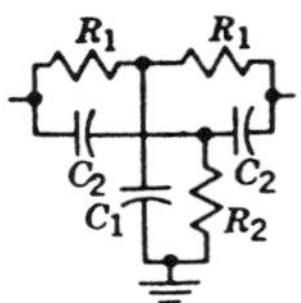 $R_1C_1 = 4R_2C_2$	$A = 2R_1$ $T_1 = \frac{R_1C_1}{2} = 2R_2C_2$ $T_2 = R_1C_2$	$R_1 = \frac{A}{2}$ $R_2 = \frac{AT_1}{4T_2}$ $C_1 = \frac{4T_1}{A}$ $C_2 = \frac{2T_2}{A}$
$\frac{1}{sB}$	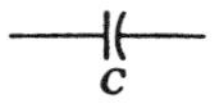	$B = C$	$C = B$
$\frac{1}{sB}(1+sT)$	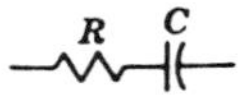	$B = C$ $T = RC$	$R = \frac{T}{B}$ $C = B$
$\frac{1}{sB}\left(\frac{1+sT}{sT}\right)$	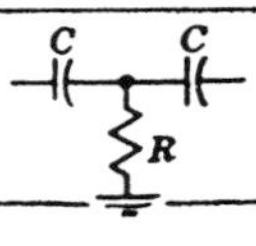	$B = \frac{C}{2}$ $T = 2RC$	$R = \frac{T}{4B}$ $C = 2B$

Transfer Functions of RC Input and Output Networks

Impedance function	Network	Relations	Inverse relations
$\frac{1}{sB}\left(\frac{1+sT}{1+s\theta T}\right)$ $\theta < 1$	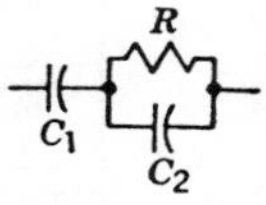	$B = C_1$ $T = R(C_1 + C_2)$ $\theta = \frac{C_2}{C_1 + C_2}$	$R = \frac{T(1-\theta)}{B}$ $C_1 = B$ $C_2 = \frac{B\theta}{1-\theta}$
	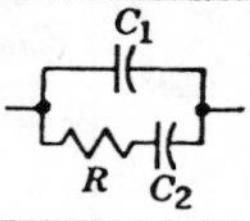	$B = C_1 + C_2$ $T = RC_2$ $\theta = \frac{C_1}{C_1 + C_2}$	$R = \frac{T}{B(1-\theta)}$ $C_1 = B\theta$ $C_2 = B(1-\theta)$
$\frac{1}{sB}\left[\frac{(1+sT_1)(1+sT_3)}{1+sT_2}\right]$ $T_1 < T_2 < T_3$	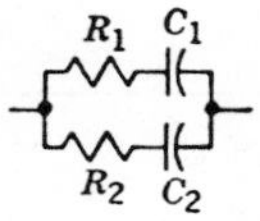	$B = C_1 + C_2$ $T_1 = R_1C_1$ $T_2 = (R_1 + R_2)\frac{C_1C_2}{C_1 + C_2}$ $T_3 = R_2C_2$	$R_1 = \frac{T_1(T_3 - T_1)}{B(T_2 - T_1)}$ $R_2 = \frac{T_3(T_3 - T_1)}{B(T_3 - T_2)}$ $C_1 = \frac{B(T_2 - T_1)}{T_3 - T_1}$ $C_2 = \frac{B(T_3 - T_2)}{T_3 - T_1}$
$A\left(\frac{1+sT_1}{1+sT_1+s^2T_1T_2}\right)$	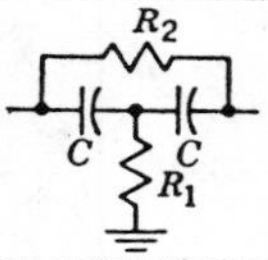	$A = R_2$ $T_1 = 2R_1C$ $T_2 = \frac{R_2C}{2}$	$R_1 = \frac{AT_1}{4T_2}$ $R_2 = A$ $C = \frac{2T_2}{A}$
$A\left(\frac{1+sT_2}{1+sT_1+s^2T_1T_2}\right)$	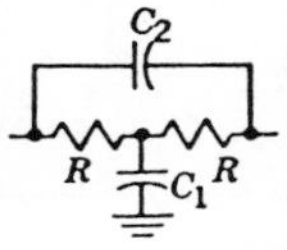	$A = 2R$ $T_1 = 2RC_2$ $T_2 = \frac{RC_1}{2}$	$R = \frac{A}{2}$ $C_1 = \frac{4T_2}{A}$ $C_2 = \frac{T_1}{A}$
$A\left[\frac{1+sT_3}{1+sT_1+s^2T_1T_2}\right]$ $T_2 > \frac{T_1}{4}$ (complex roots) $T_3 > T_2$	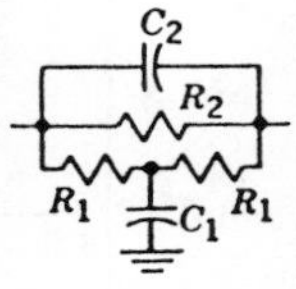	$A = \frac{2R_1R_2}{2R_1 + R_2}$ $T_1 = \frac{R_1(R_1C_1 + 2R_2C_2)}{2R_1 + R_2}$ $T_2 = \frac{R_1R_2C_1C_2}{R_1C_1 + 2R_2C_2}$ $T_3 = \frac{R_1C_1}{2}$	$R_1 = \frac{AT_3^2}{2[T_3^2 - T_1(T_3 - T_2)]}$ $R_2 = \frac{AT_3^2}{T_1(T_3 - T_2)}$ $C_1 = \frac{4[T_3^2 - T_1(T_3 - T_2)]}{AT_3}$ $C_2 = \frac{T_1T_2}{AT_3}$
	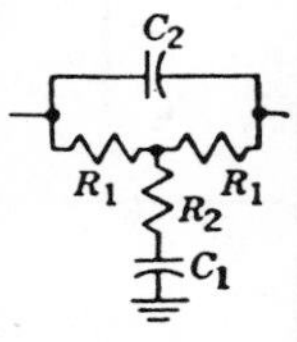	$A = 2R_1$ $T_1 = R_2C_1 + 2R_1C_2$ $T_2 = \frac{R_1(R_1 + 2R_2)C_1C_2}{R_2C_1 + 2R_1C_2}$ $T_3 = \left(R_2 + \frac{R_1}{2}\right)C_1$	$R_1 = \frac{A}{2}$ $R_2 = \frac{AT_1(T_3 - T_2)}{4[T_3^2 - T_1(T_3 - T_2)]}$ $C_1 = \frac{4[T_3^2 - T_1(T_3 - T_2)]}{AT_3}$ $C_2 = \frac{T_1T_2}{AT_3}$
	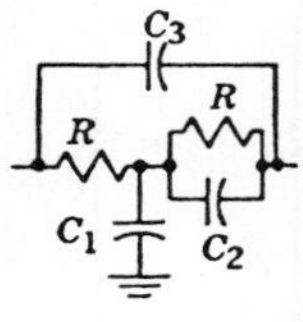	$A = 2R$ $T_1 = R(C_2 + 2C_3)$ $T_2 = \frac{RC_3(C_1 + C_2)}{C_2 + 2C_3}$ $T_3 = \frac{R}{2}(C_1 + C_2)$	$R = \frac{A}{2}$ $C_1 = \frac{2[2T_3^2 - T_1(T_3 - T_2)]}{AT_3}$ $C_2 = \frac{2T_1(T_3 - T_2)}{AT_3}$ $C_3 = \frac{T_1T_2}{AT_3}$

Transfer-function Plots For Typical Transfer Function

	$G(s)$	Polar plot	Bode diagram
1.	$\dfrac{K}{s\tau_1 + 1}$	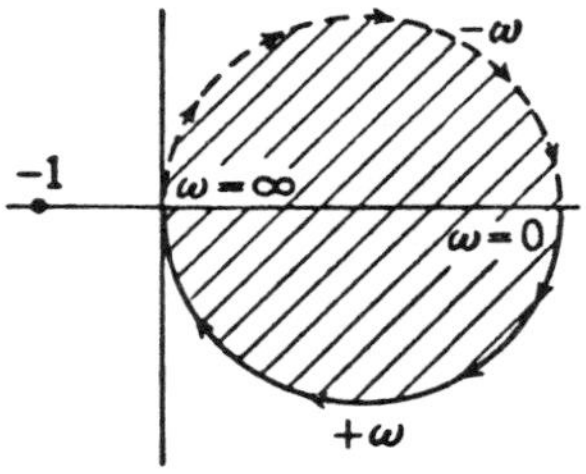	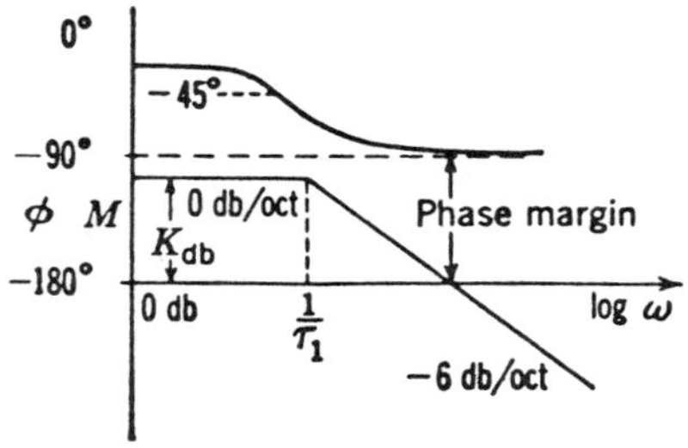
2.	$\dfrac{K}{(s\tau_1 + 1)(s\tau_2 + 1)}$	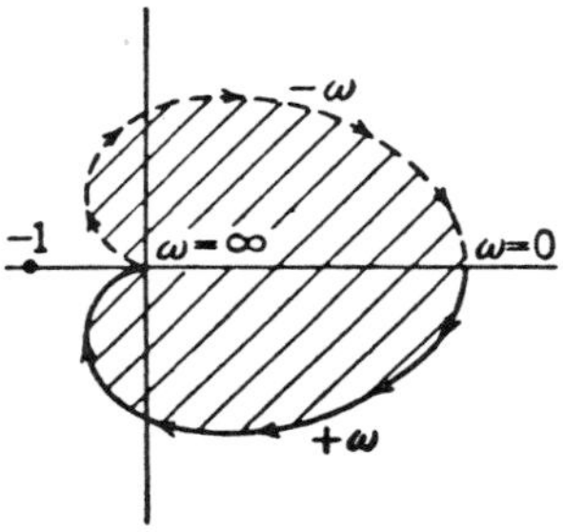	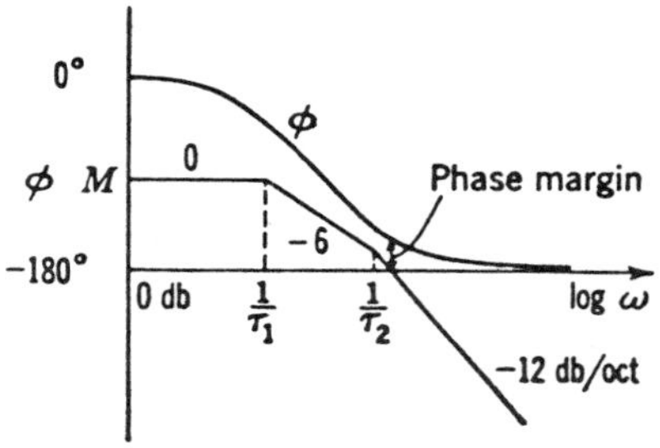
3.	$\dfrac{K}{(s\tau_1 + 1)(s\tau_2 + 1)(s\tau_3 + 1)}$	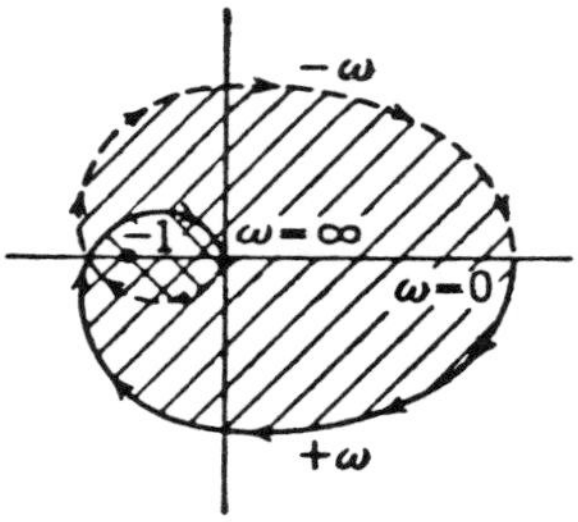	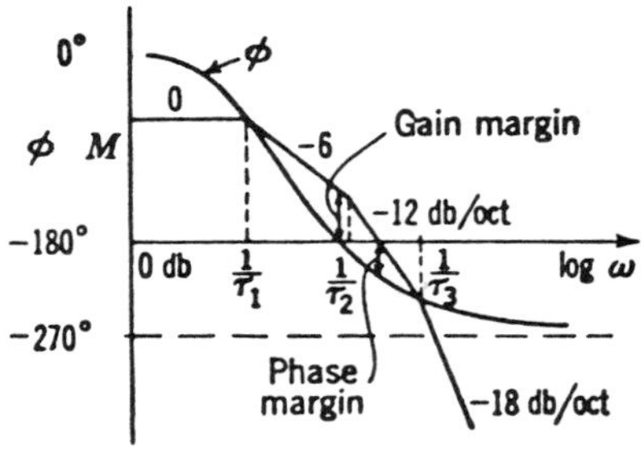
4.	$\dfrac{K}{s}$	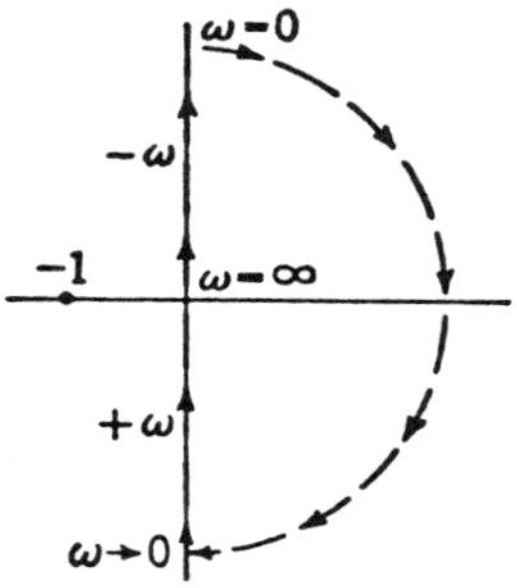	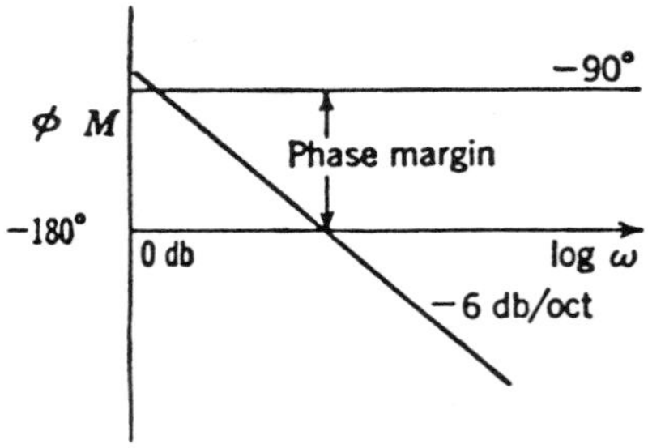

<table>
<tr><th>Nichols diagram</th><th>Root locus</th><th>Comments</th></tr>
<tr><td>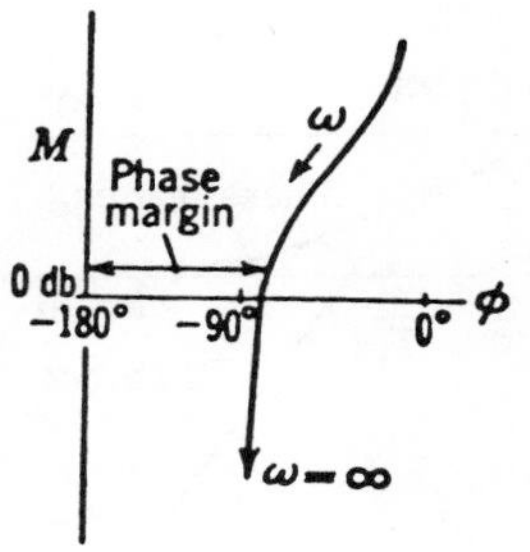</td><td>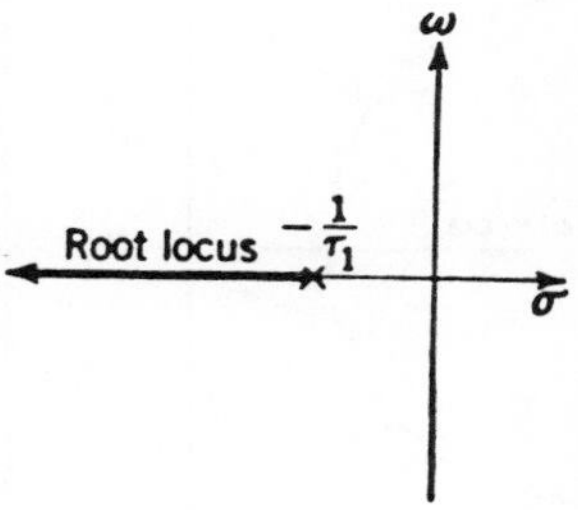</td><td>Stable; gain margin = ∞</td></tr>
<tr><td>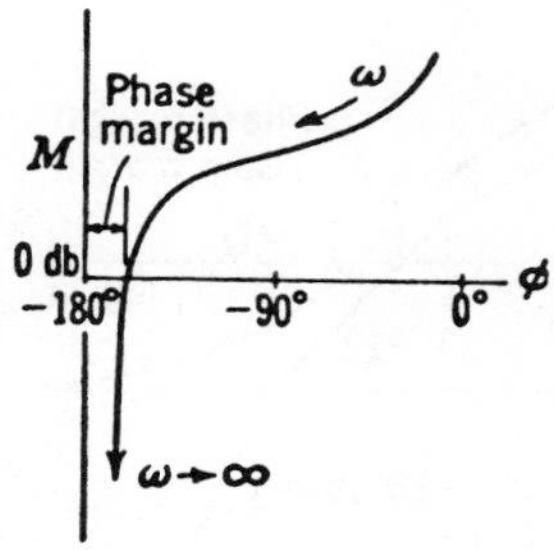</td><td>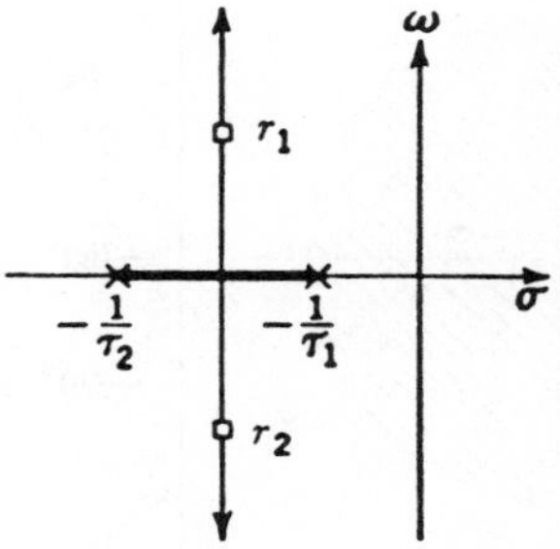</td><td>Elementary regulator; stable; gain margin = ∞</td></tr>
<tr><td>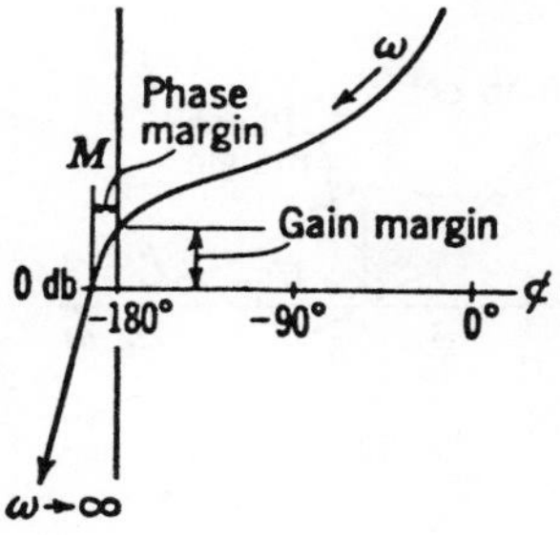</td><td>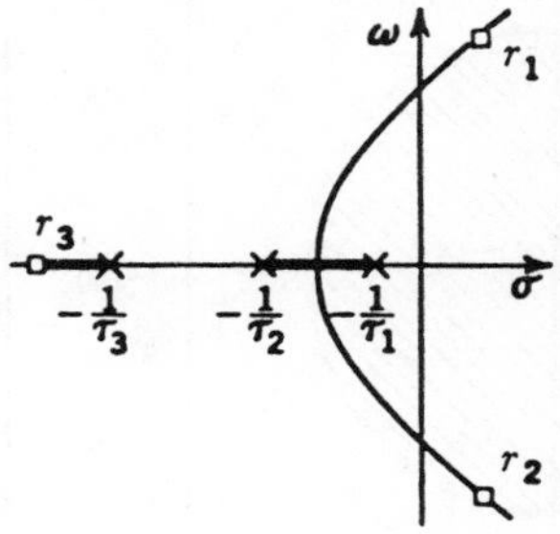</td><td>Regulator with additional energy-storage component; unstable, but can be made stable by reducing gain</td></tr>
<tr><td>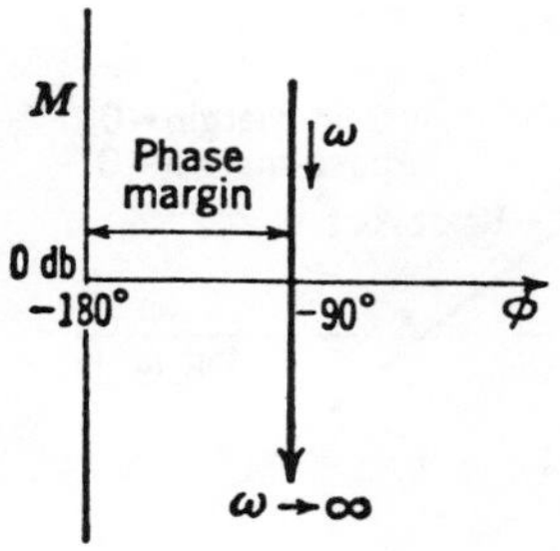</td><td>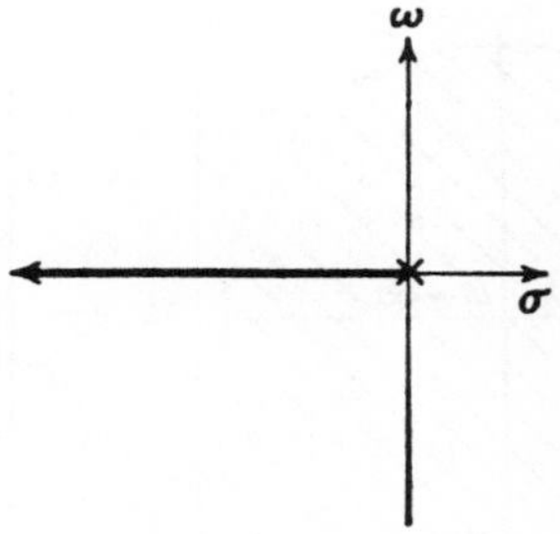</td><td>Ideal integrator; stable</td></tr>
</table>

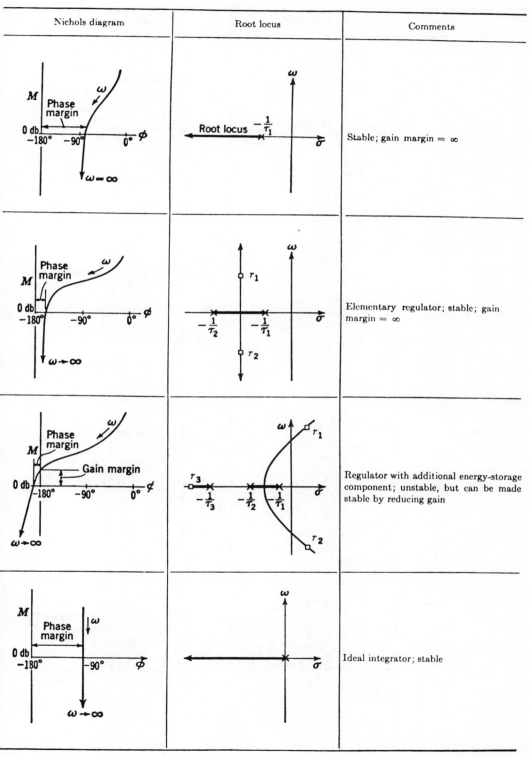

(*Continued*)

(*Continued*)

<table>
<tr><th>G(s)</th><th>Polar plot</th><th>Bode diagram</th></tr>
<tr><td>5. $\dfrac{K}{s(s\tau_1 + 1)}$</td><td>
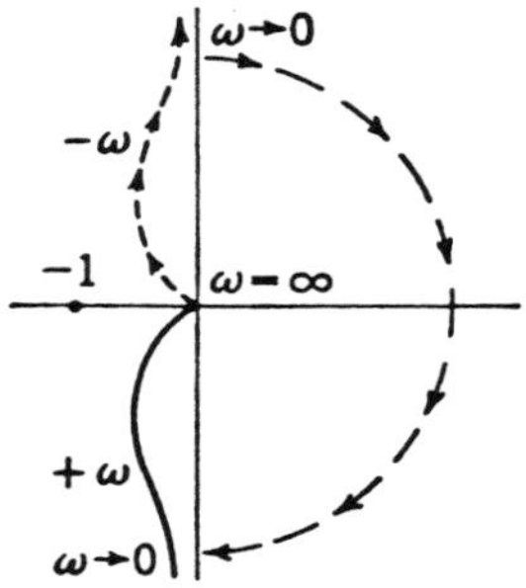

</td><td>
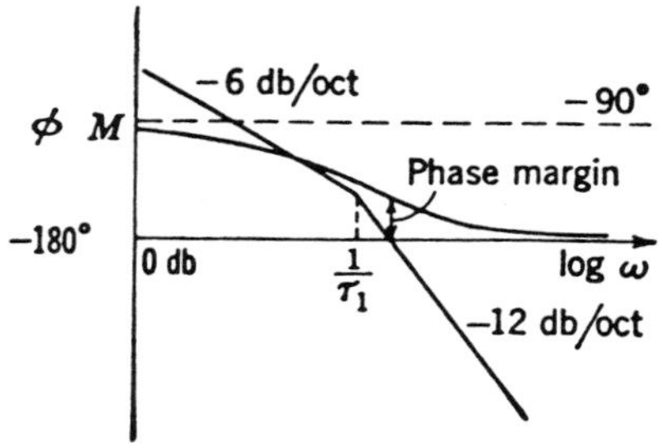

</td></tr>
<tr><td>6. $\dfrac{K}{s(s\tau_1 + 1)(s\tau_2 + 1)}$</td><td>
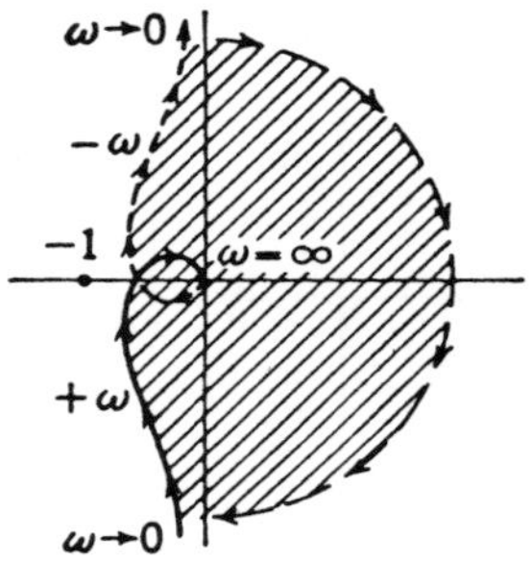

</td><td>
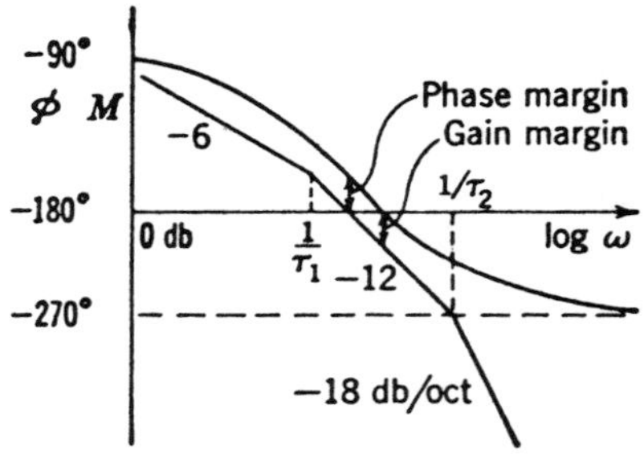

</td></tr>
<tr><td>7. $\dfrac{K(s\tau_a + 1)}{s(s\tau_1 + 1)(s\tau_2 + 1)}$</td><td>
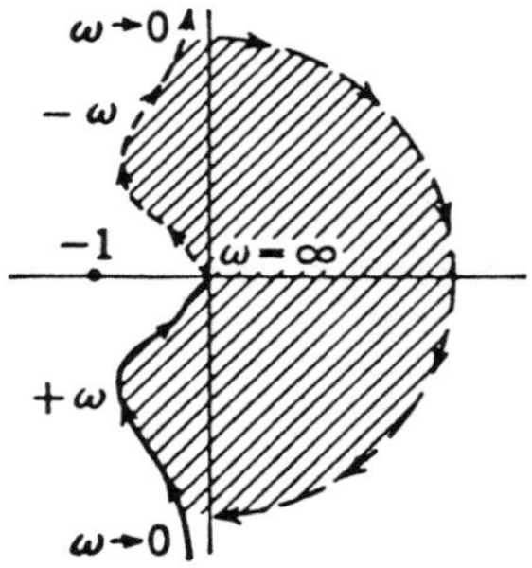

</td><td>
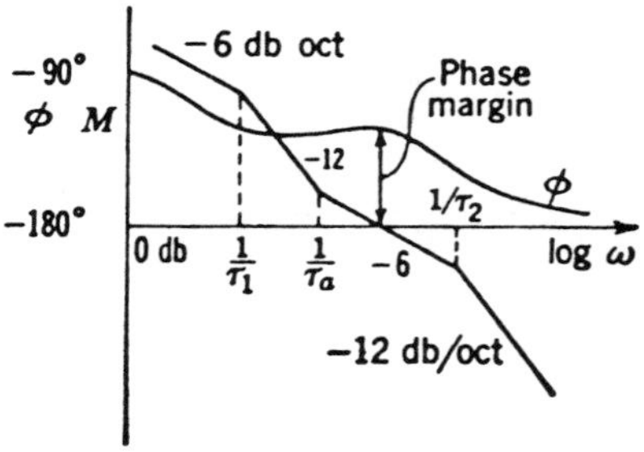

</td></tr>
<tr><td>8. $\dfrac{K}{s^2}$</td><td>
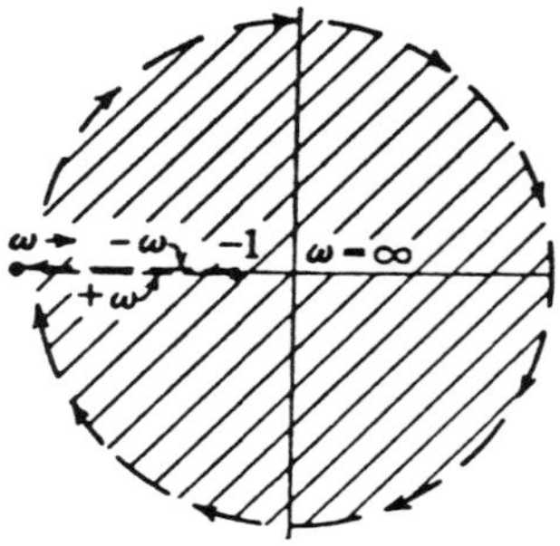

</td><td>
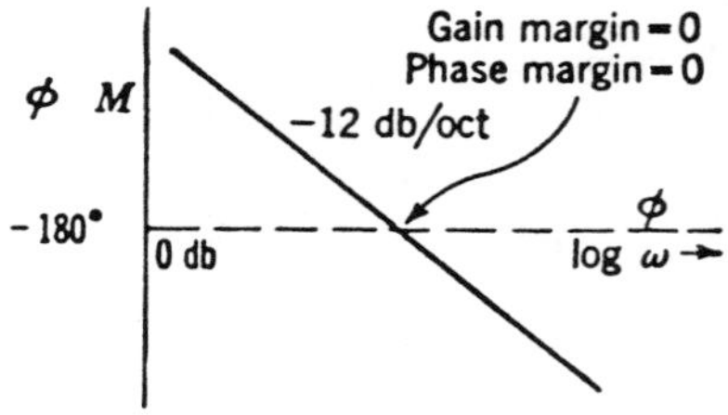

</td></tr>
</table>

Nichols diagram	Root locus	Comments
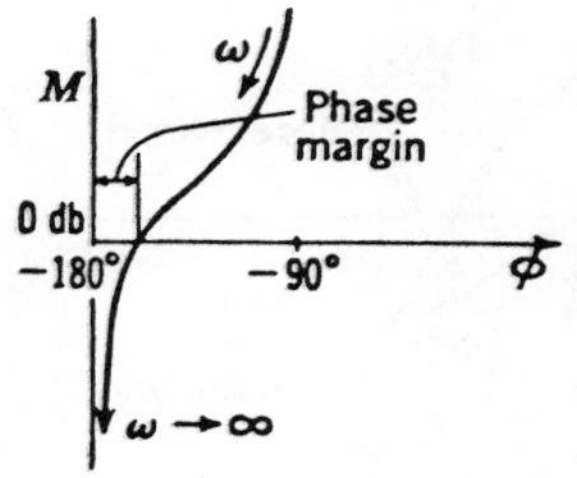	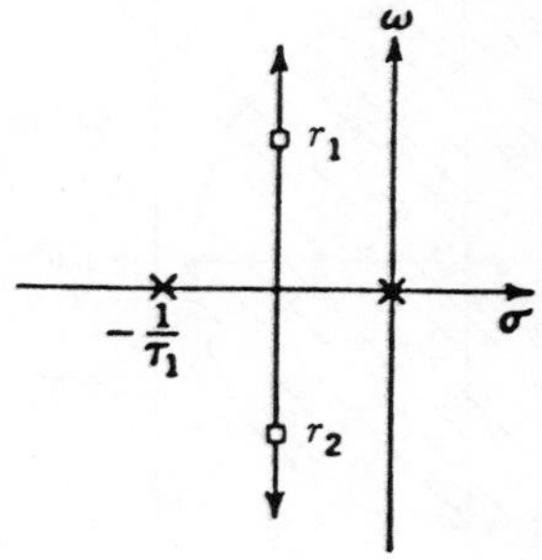	Elementary instrument servo; inherently stable; gain margin = ∞
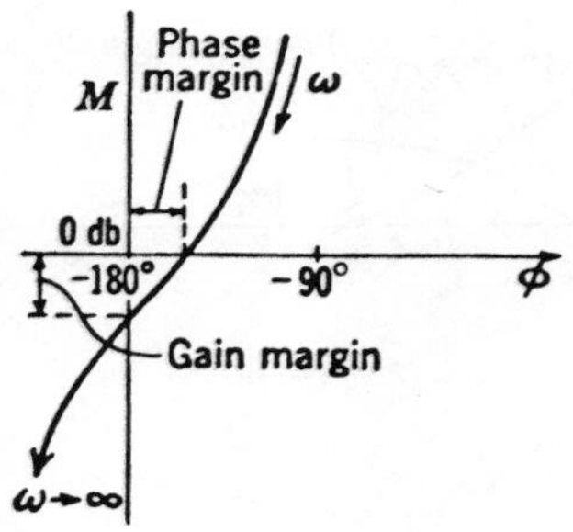	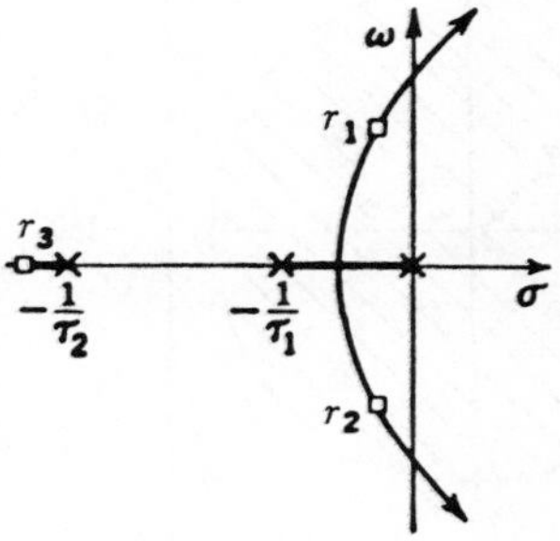	Instrument servo with field-control motor or power servo with elementary Ward-Leonard drive; stable as shown, but may become unstable with increased gain
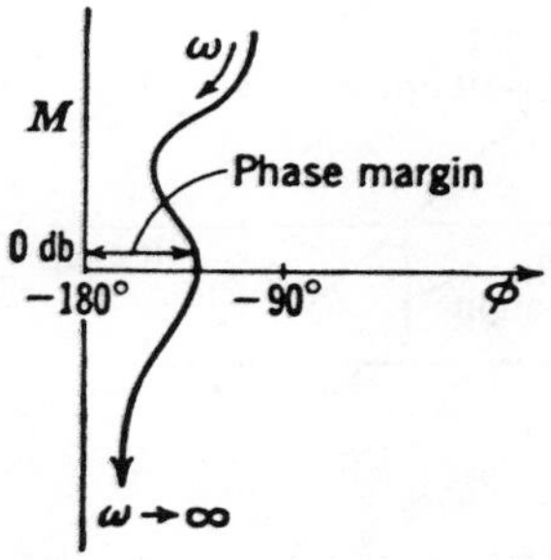	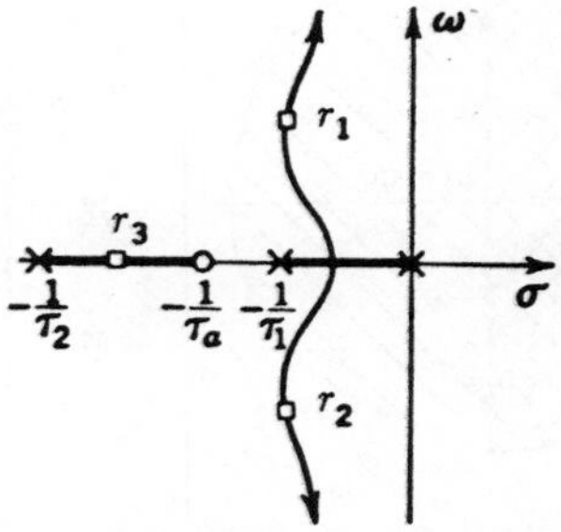	Elementary instrument servo with phase-lead (derivative) compensator; stable
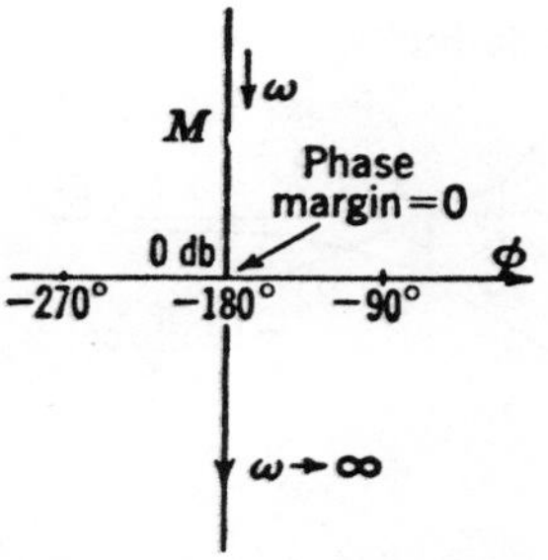	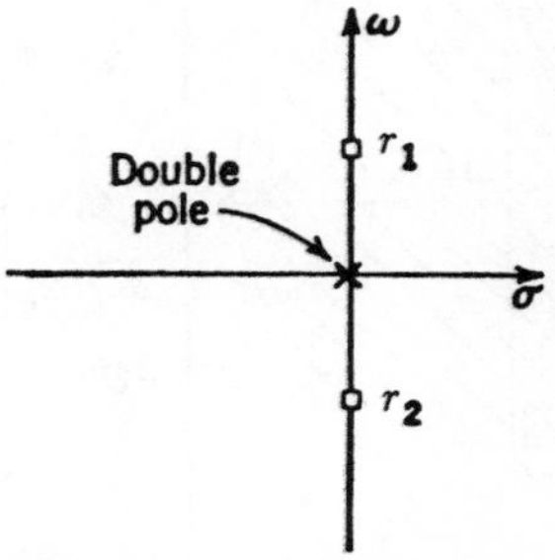	Inherently unstable; must be compensated

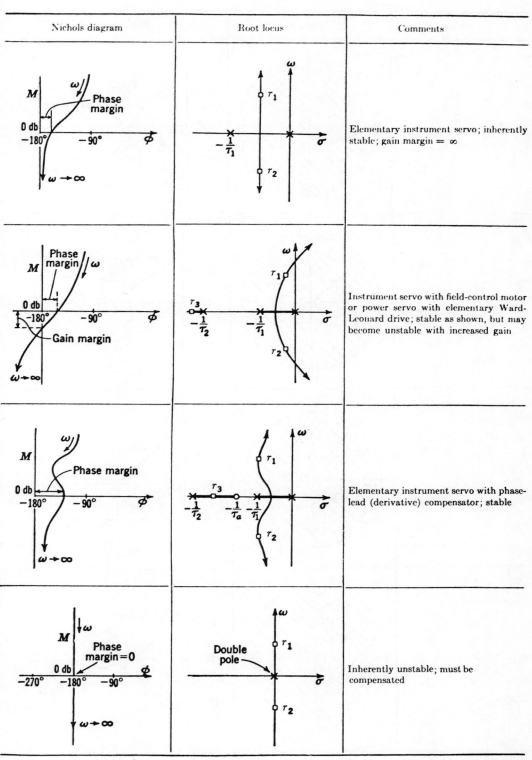

(Continued)

(*Continued*)

	$G(s)$	Polar plot	Bode diagram
9.	$\frac{K}{s^2(s\tau_1+1)}$	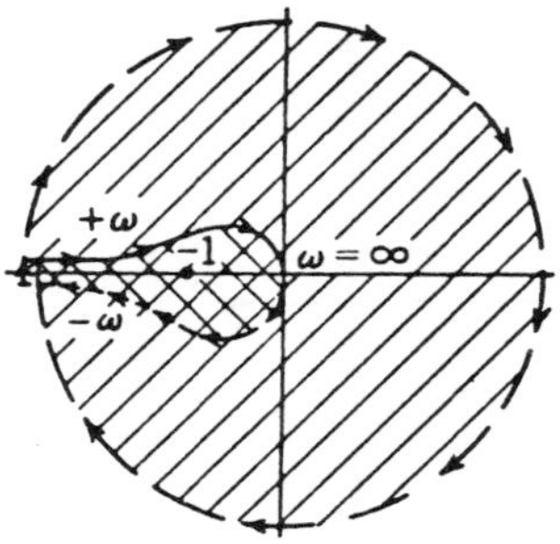	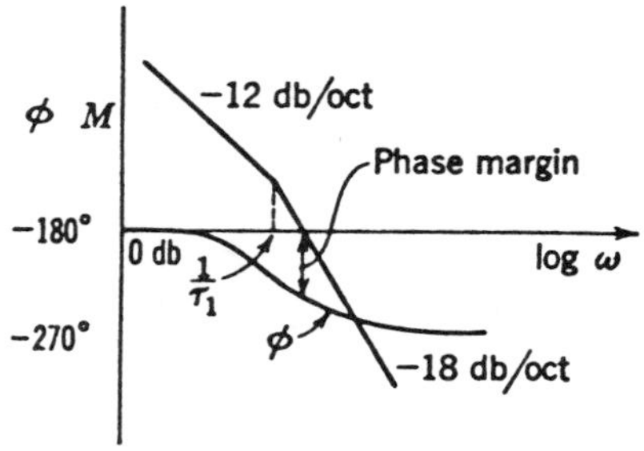
10.	$\frac{K(s\tau_a+1)}{s^2(s\tau_1+1)}$ $\tau_a > \tau_1$	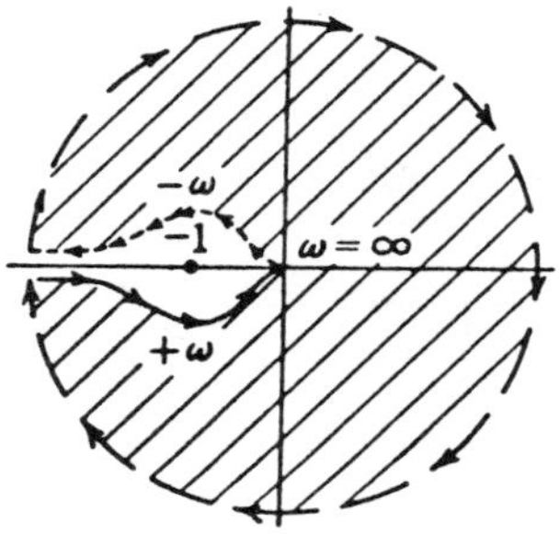	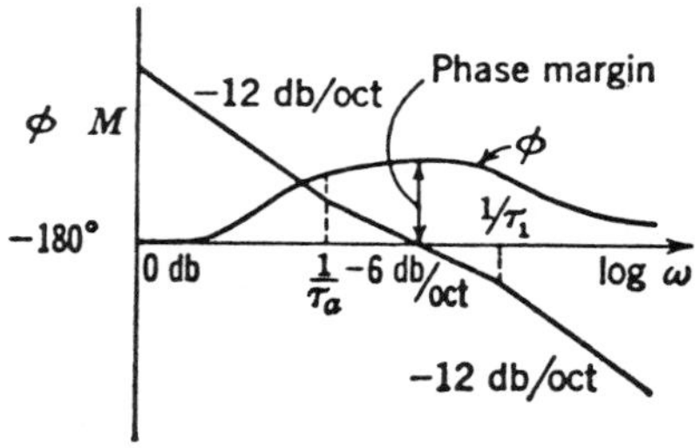
11.	$\frac{K}{s^3}$	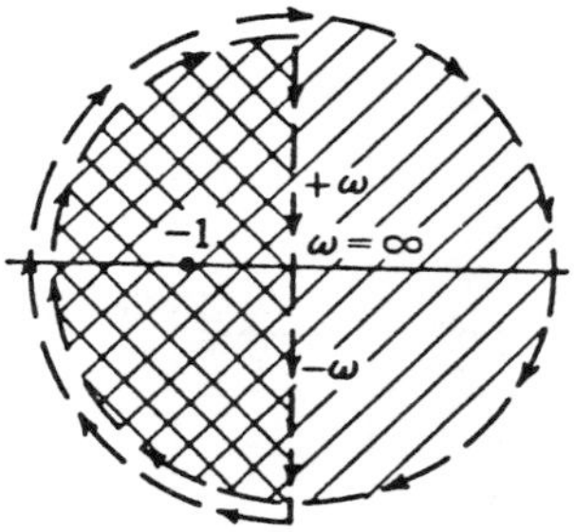	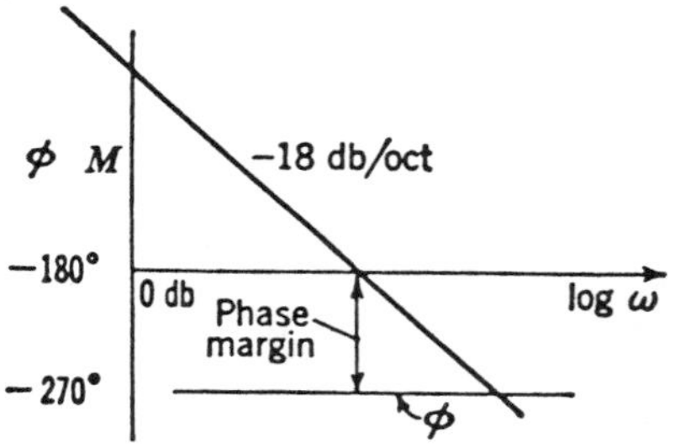
12.	$\frac{K(s\tau_a+1)}{s^3}$	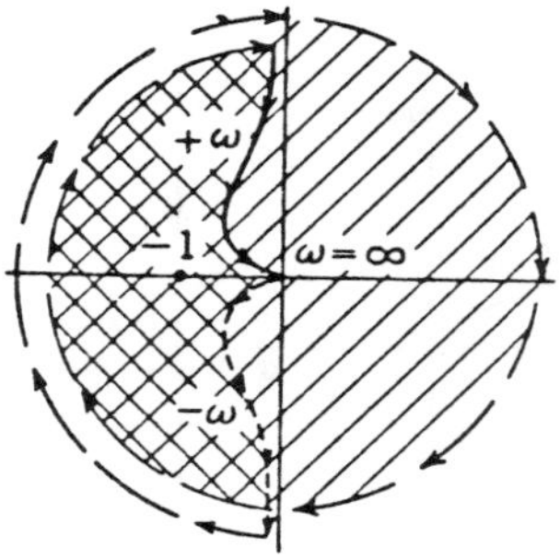	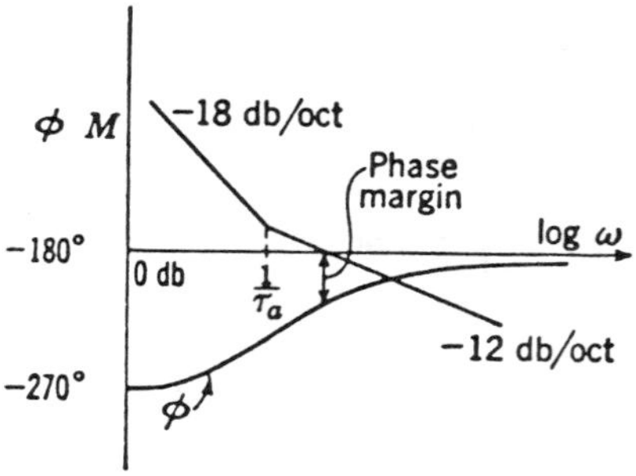

Nichols diagram	Root locus	Comments
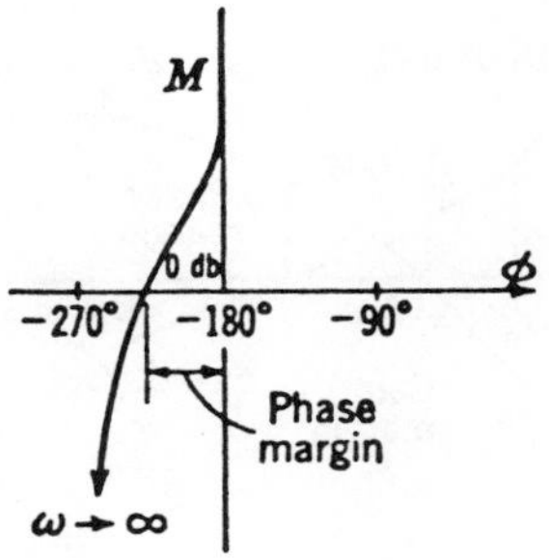	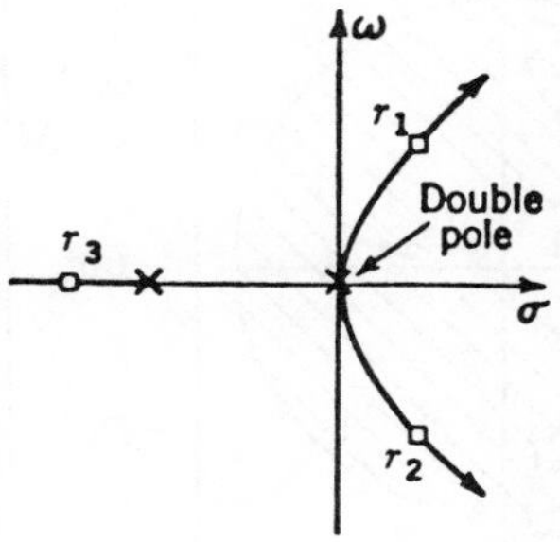	Inherently unstable; must be compensated
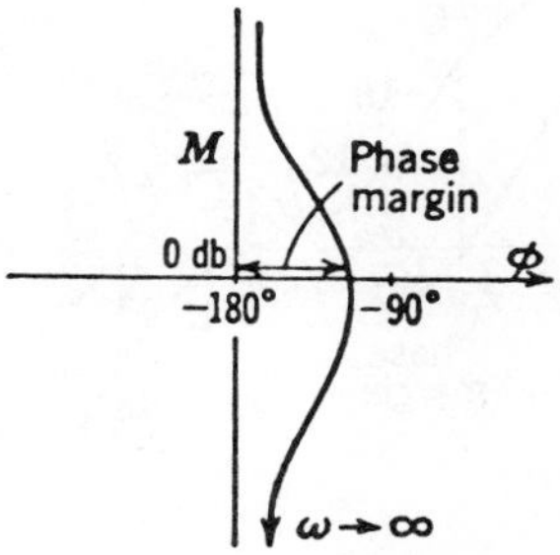	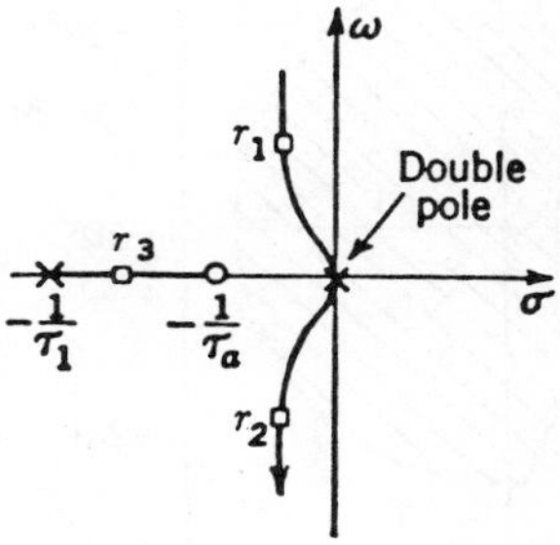	Stable for all gains
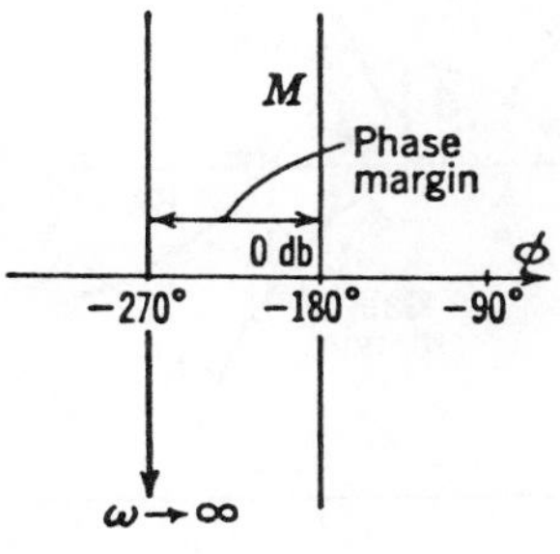	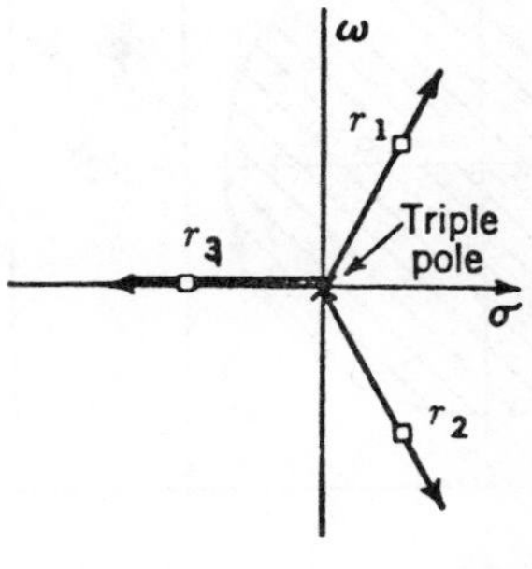	Inherently unstable
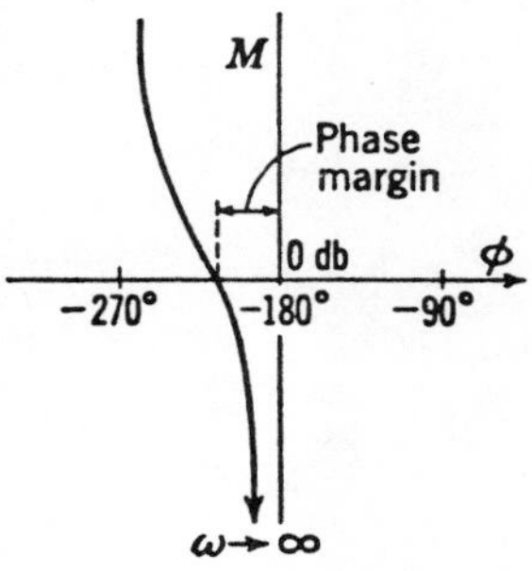	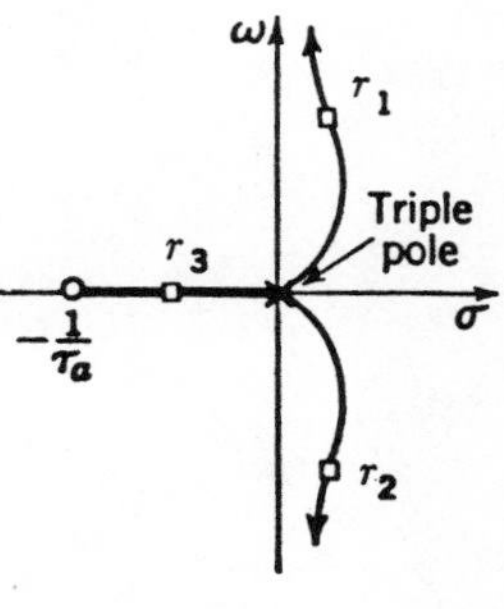	Inherently unstable

(Continued)

(Continued)

$G(s)$	Polar plot	Bode diagram
13. $\dfrac{K(s\tau_a+1)(s\tau_b+1)}{s^3}$	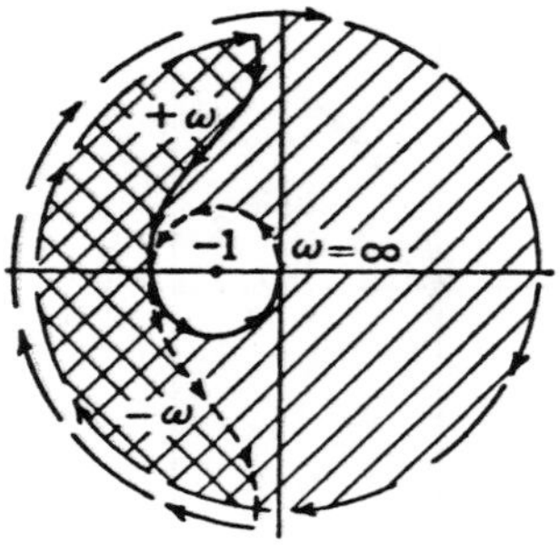	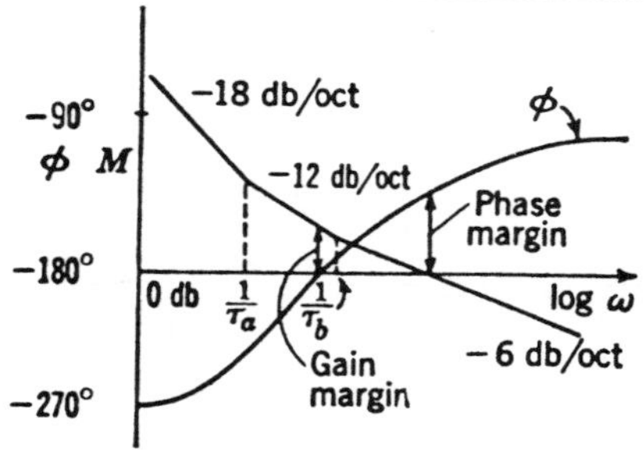
14. $\dfrac{K(s\tau_a+1)(s\tau_b+1)}{s(s\tau_1+1)(s\tau_2+1)(s\tau_3+1)(s\tau_4+1)}$	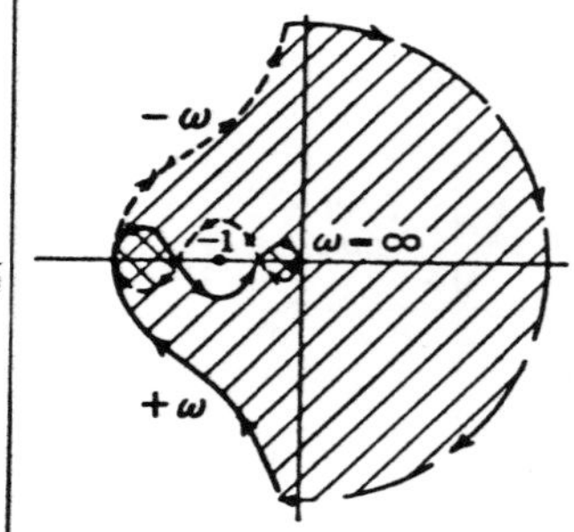	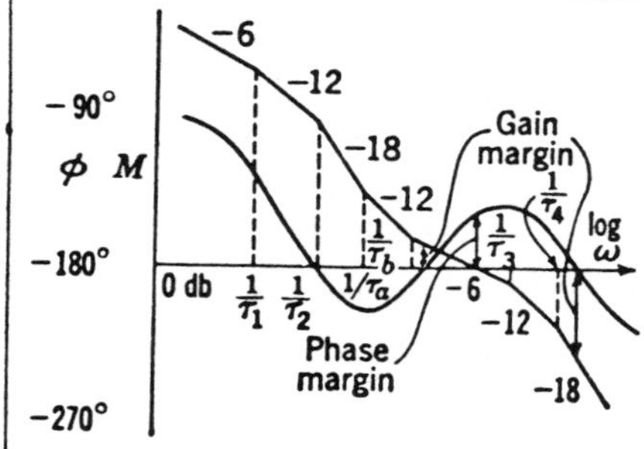
15. $\dfrac{K(s\tau_a+1)}{s^2(s\tau_1+1)(s\tau_2+1)}$	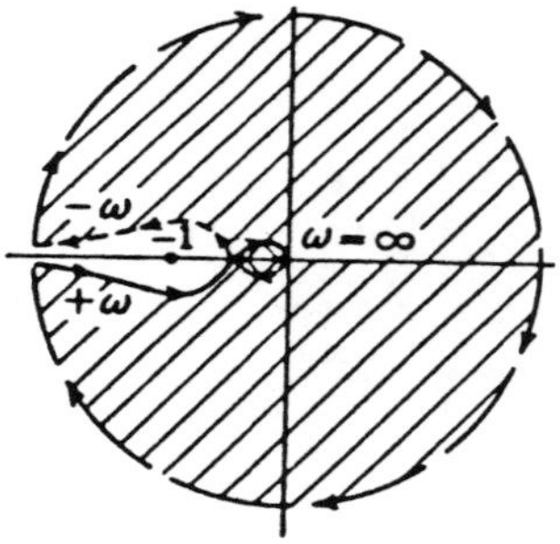	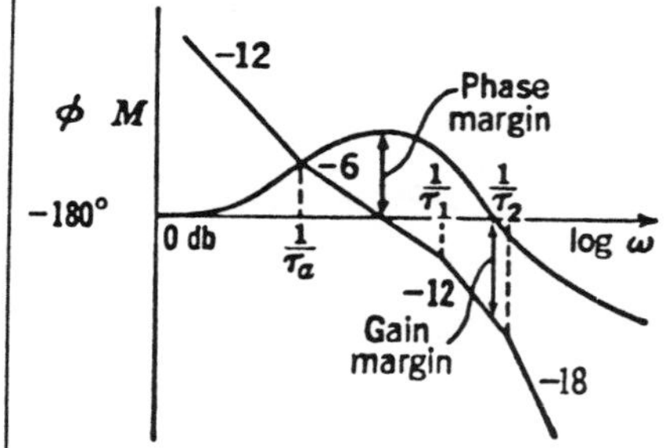

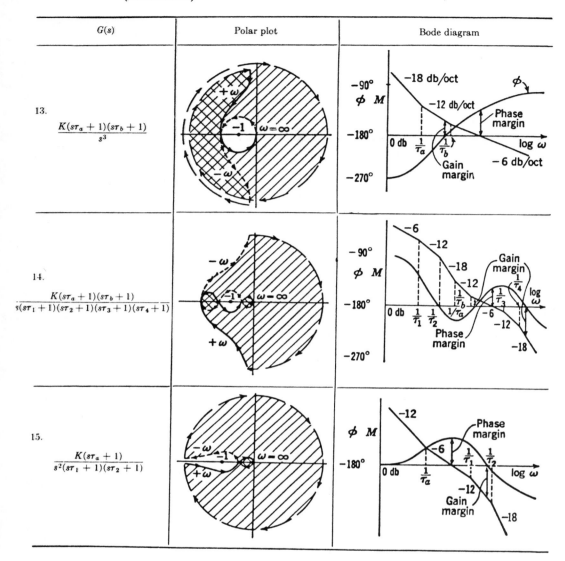

<table>
<tr><th>Nichols diagram</th><th>Root locus</th><th>Comments</th></tr>
<tr><td>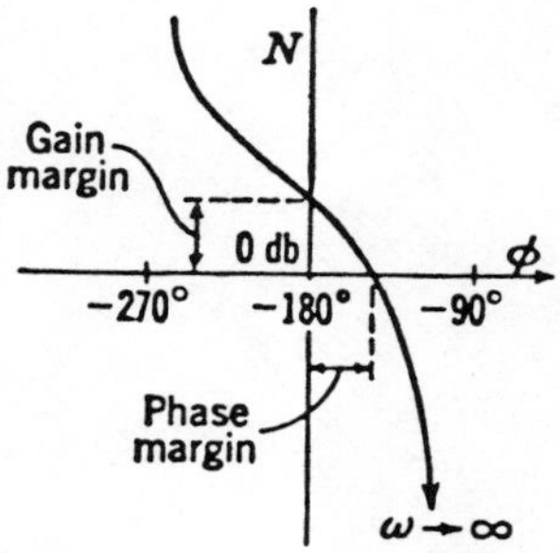</td><td>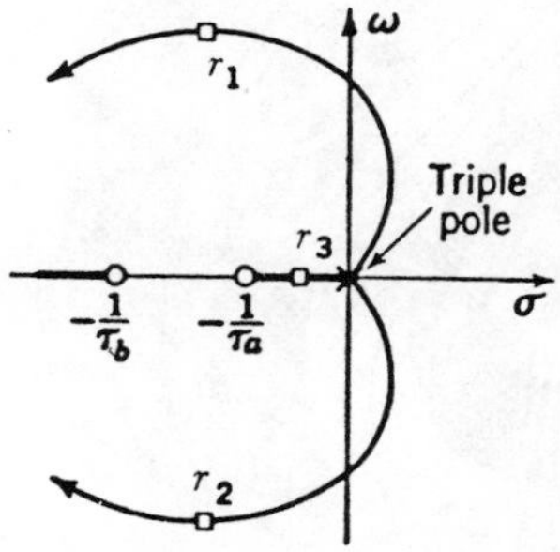</td><td>Conditionally stable; becomes unstable if gain is too low</td></tr>
<tr><td>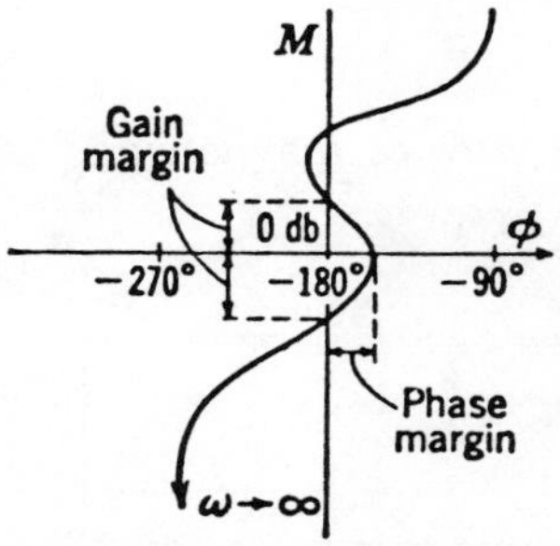</td><td>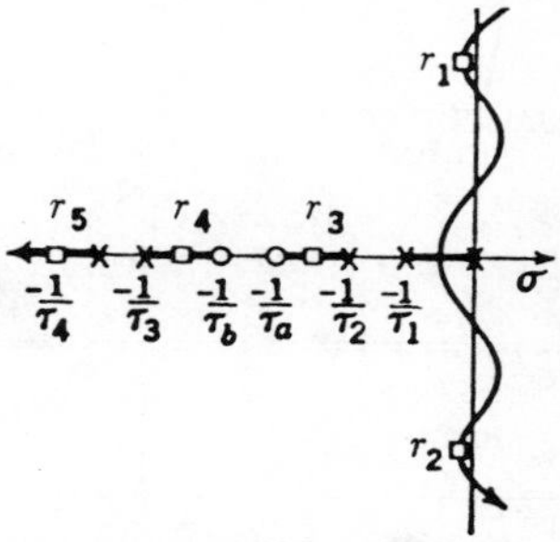</td><td>Conditionally stable; stable at low gain, becomes unstable as gain is raised, again becomes stable as gain is further increased, and becomes unstable for very high gains</td></tr>
<tr><td>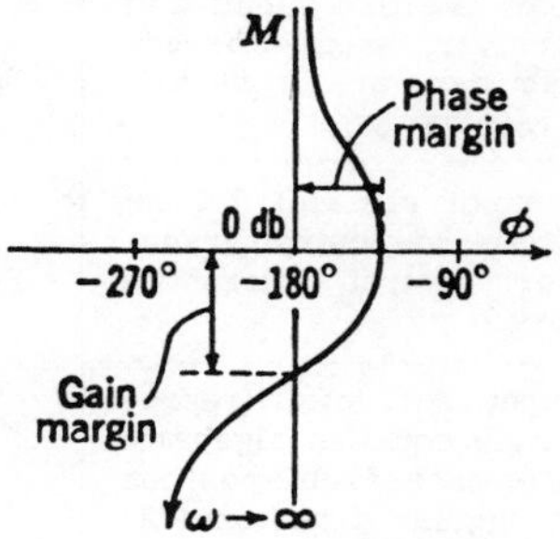</td><td>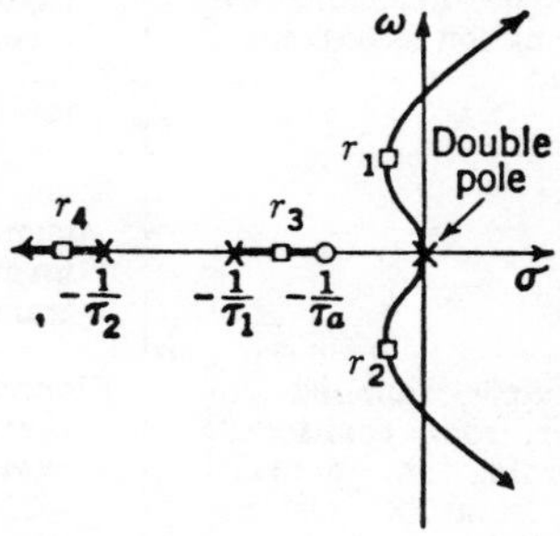</td><td>Conditionally stable; becomes unstable at high gain</td></tr>
</table>

Appendix H: DEVICES USED IN AUTOMATIC CONTROL

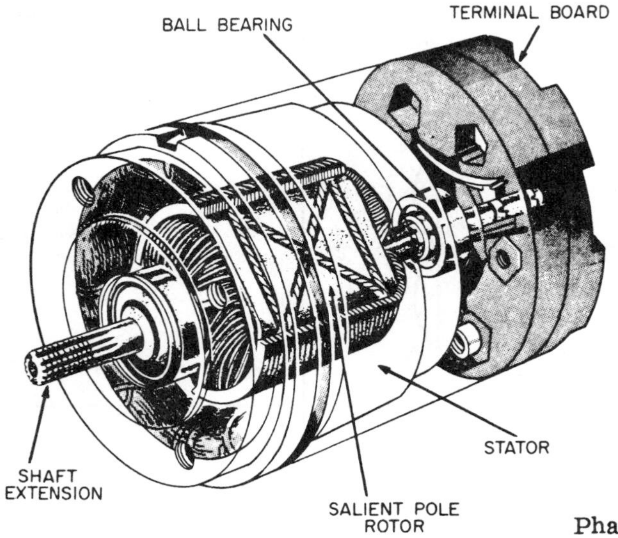

Phantom view of a synchro.

Synchro information

Functional Classification	Military Abbre-viations	Input	Output
Torque Trans-mitter	TX	Rotor positioned mechanically or manually by information to be transmitted	Electrical output from stator identifying rotor position supplied to torque receiver, torque differential transmitter, or torque differential receiver
Control Trans-mitter	CX	Same as TX	Electrical output same as TX but supplied only to control transformer or control differential transmitter
Torque Differential Transmitter	TDX	TX output applied to stator; rotor positioned according to amount data from TX must be modified	Electric output from rotor (representing angle equal to algebraic sum or difference of rotor position angle and angular data from TX) supplied to torque receivers, another TDX, or a torque differential receiver
Control Differential Transmitter	CDX	Same as TDX but data usually supplied by CX	Same as TDX but supplied only to control transformer or another CDX
Torque Receiver	TR	Electrical angular position data from TX or TDX supplied to stator	Rotor assumes position determined by electrical input supplied
Torque Differential Receiver	TDR	Electrical data supplied from two TDX's, two TX's or one TX and one TDX (one connected to rotor, one to stator)	Rotor assumes position equal to algebraic sum or difference of two angular inputs
Control Trans-former	CT	Electrical data from CX or CDX applied to stator; rotor positioned mechanically or manually	Electrical output from rotor (proportional to sine of the difference between rotor angular position and electrical input angle)

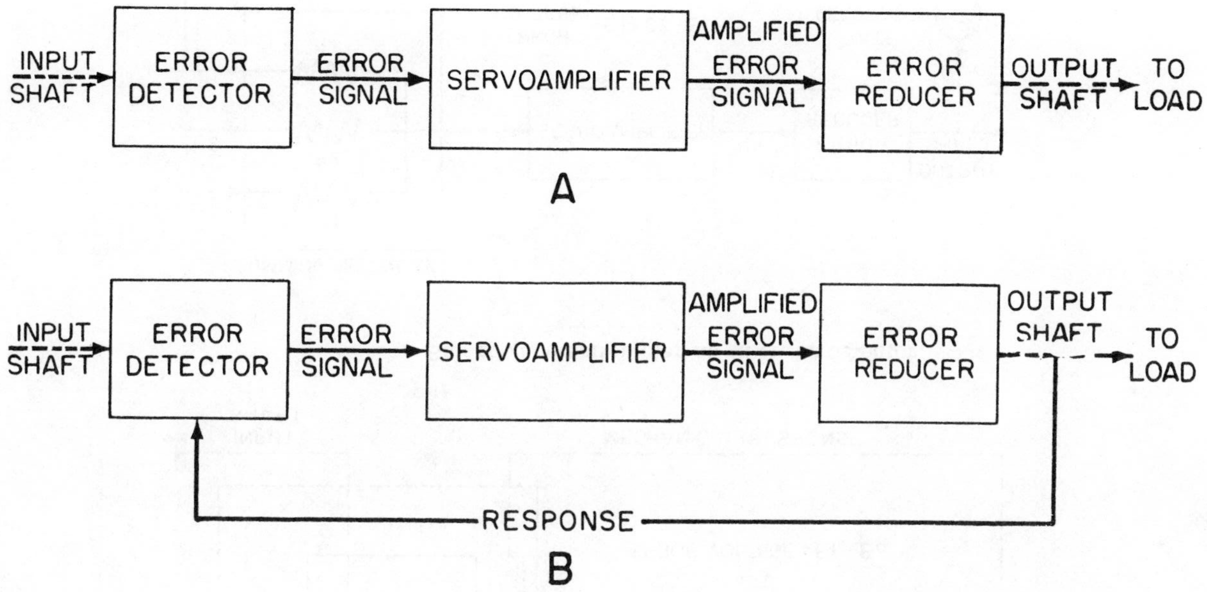

Basic servosystems: (A) Open loop; (B) Closed loop.

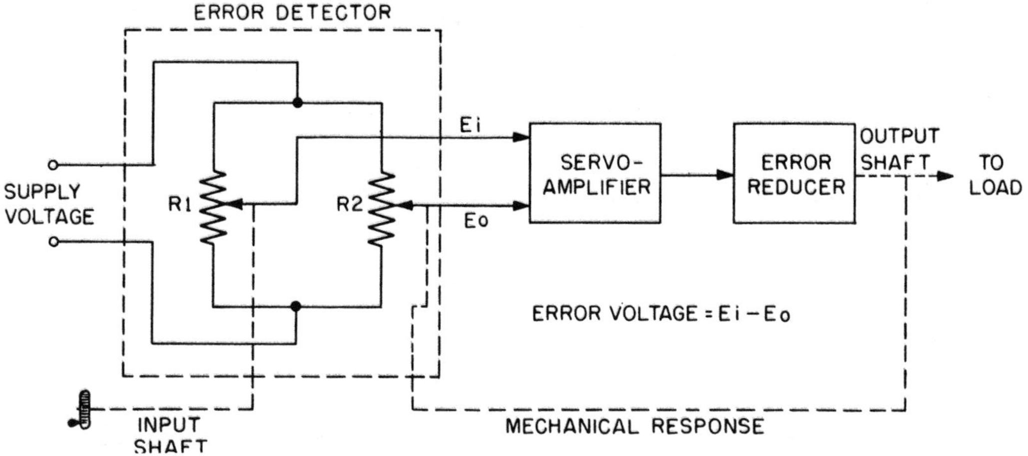

Balanced potentiometer as an error detector.

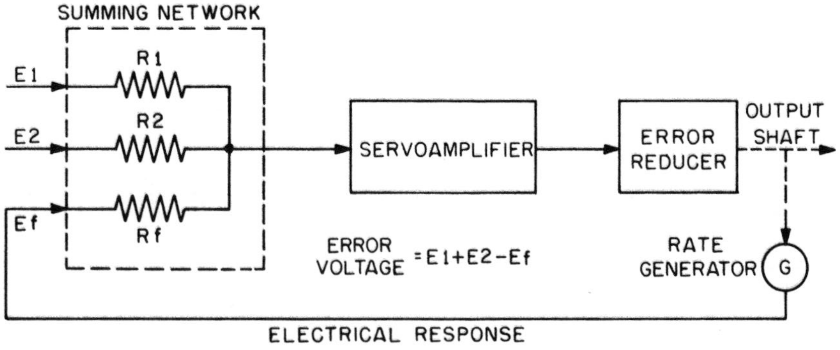

Summing network as an error detector.

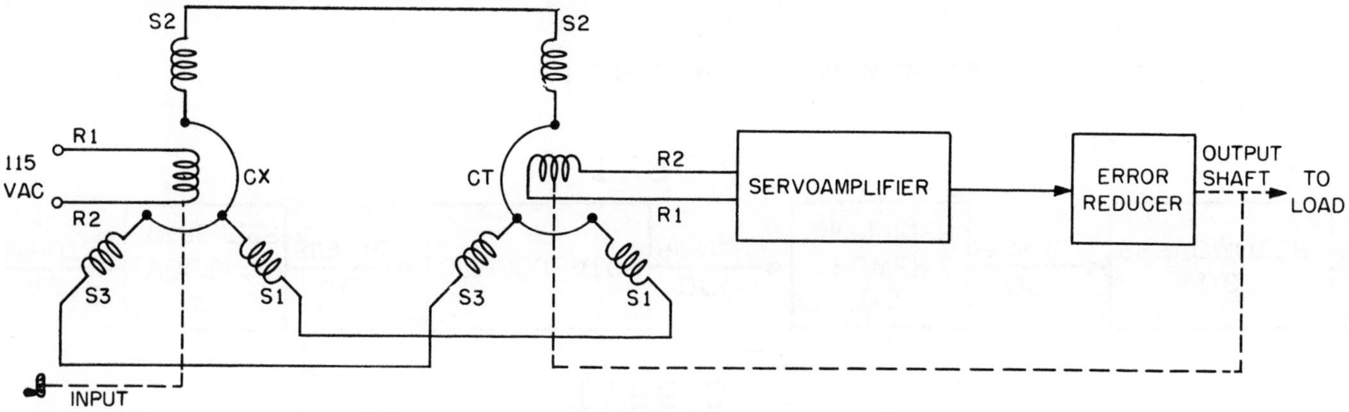

The CT as an error detector.

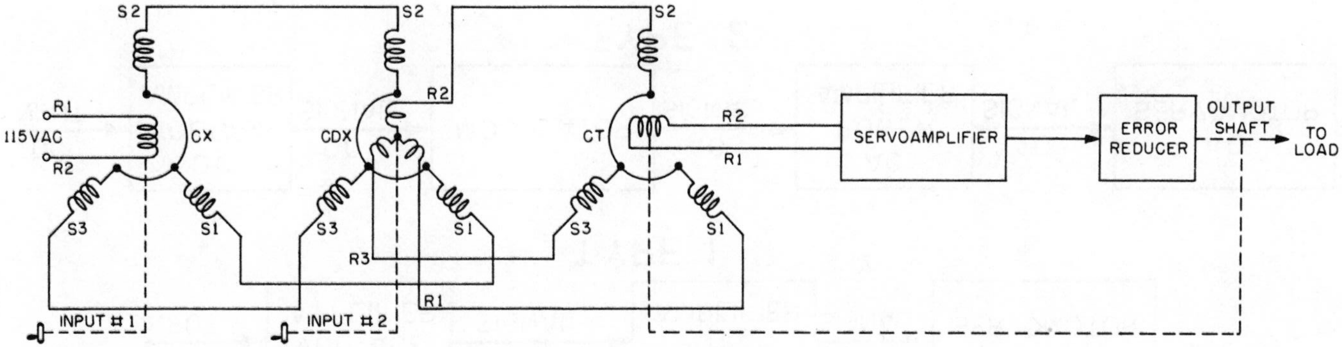

The CDX as an additional input.

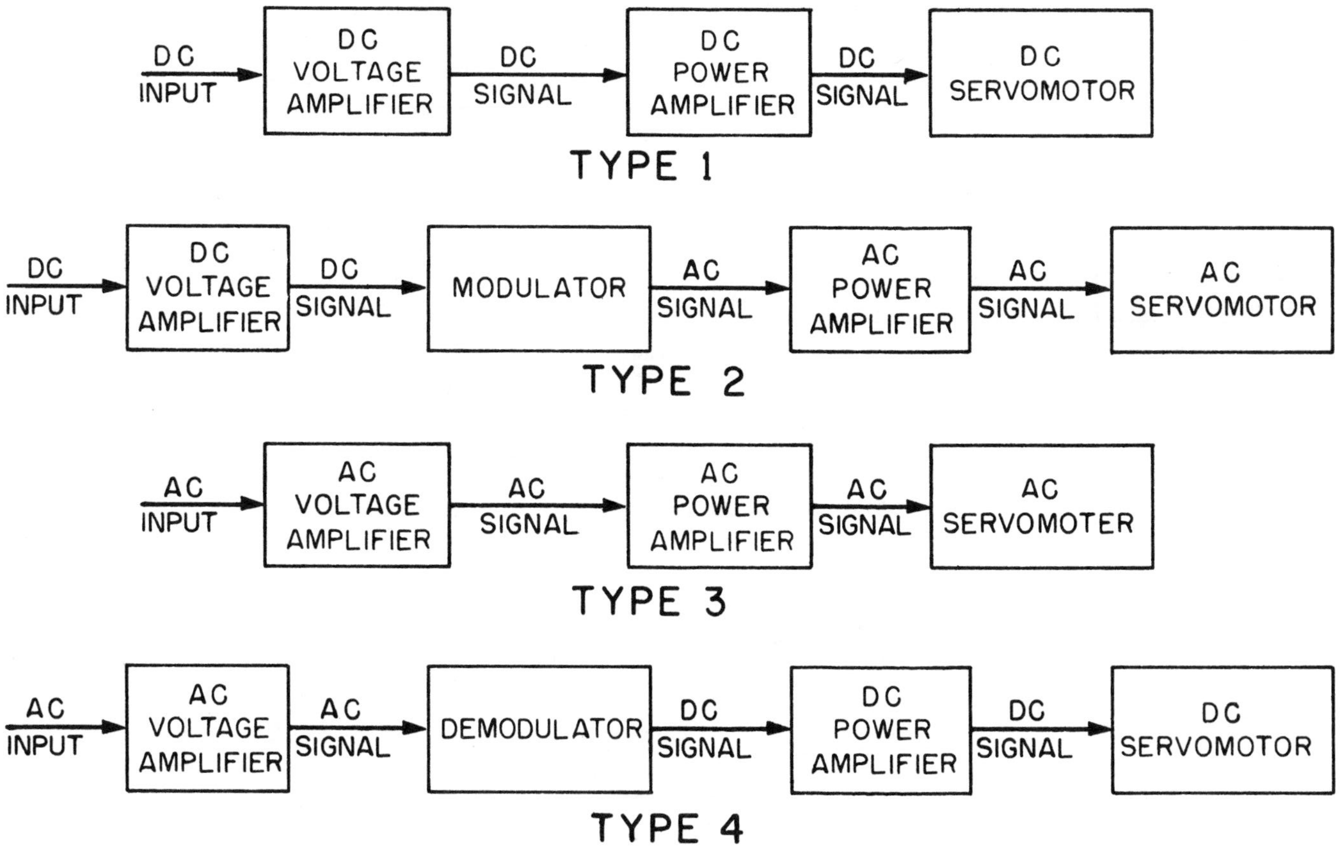

Basic types of servoamplifiers.

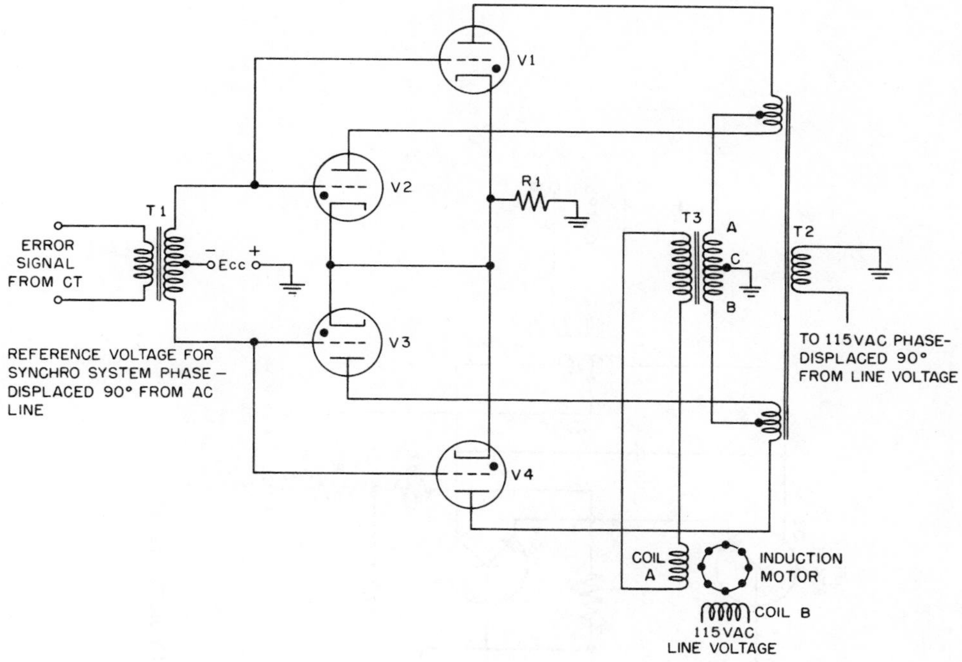

Thyratron servoamplifier.

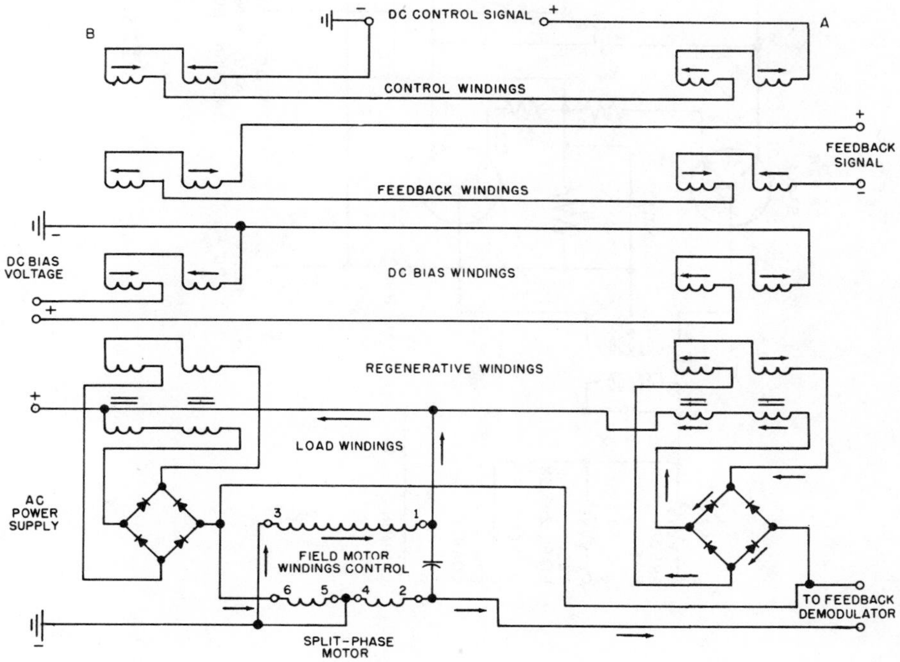

Magnetic servoamplifier.

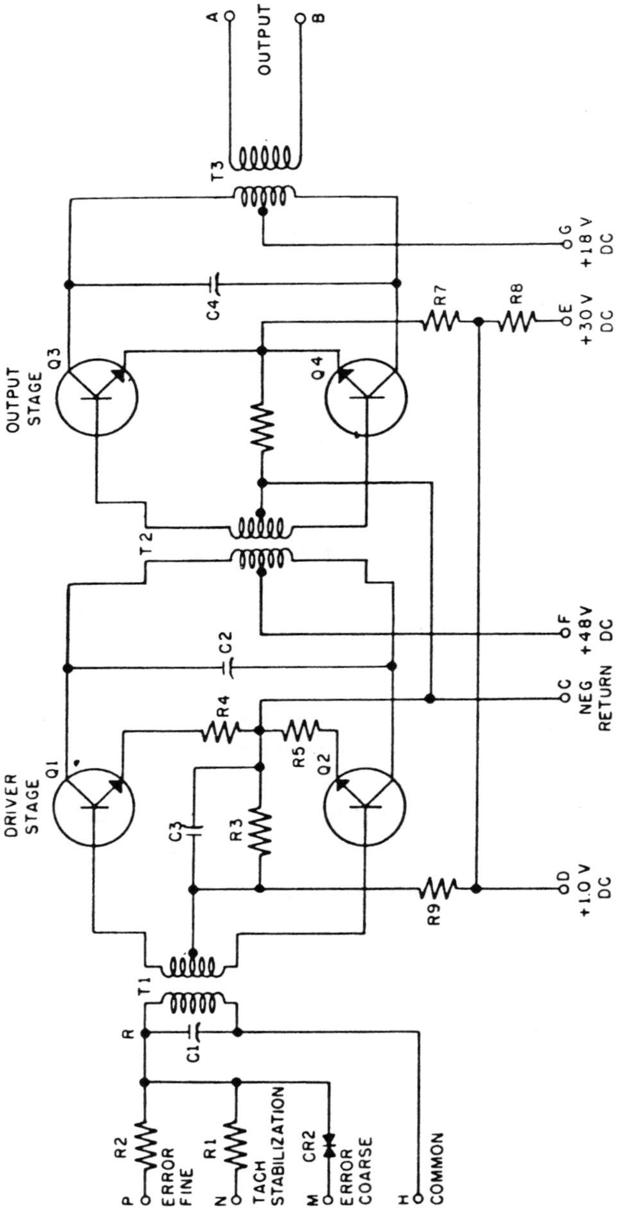

Transistor servoamplifier.

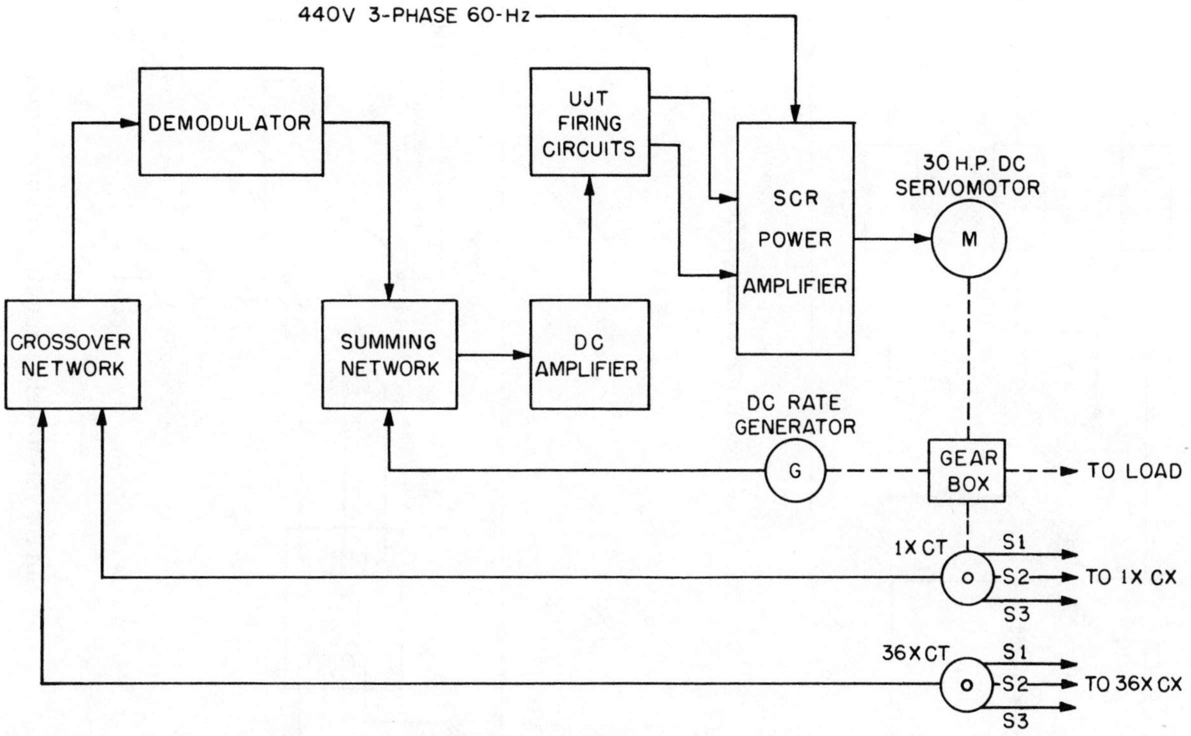

Servo using SCR power amplifier.

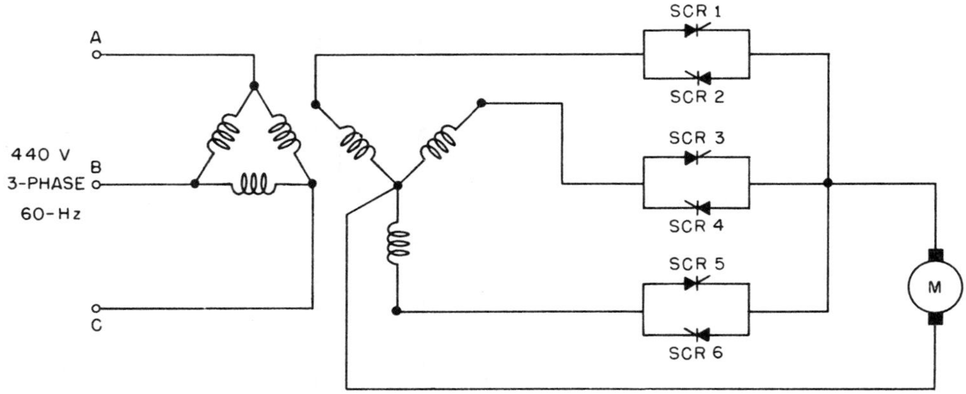

SCR power amplifier.

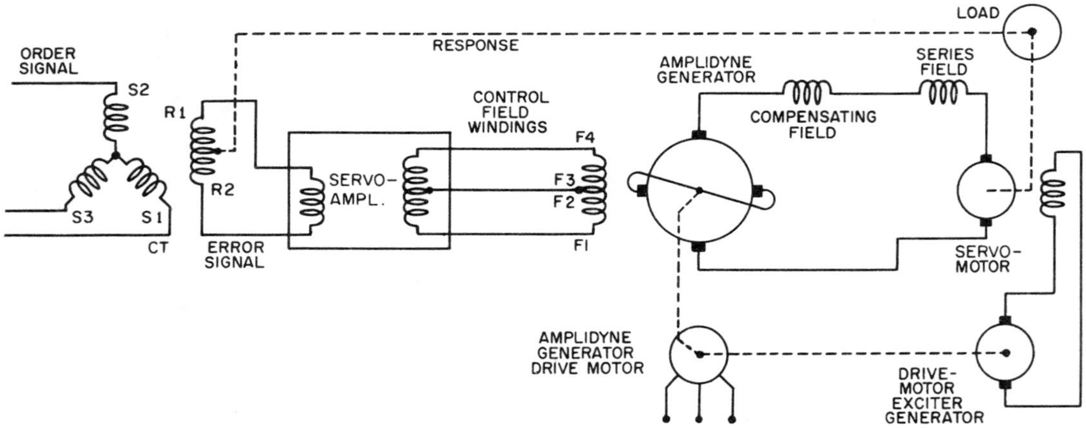

Servosystem using amplidyne generator.

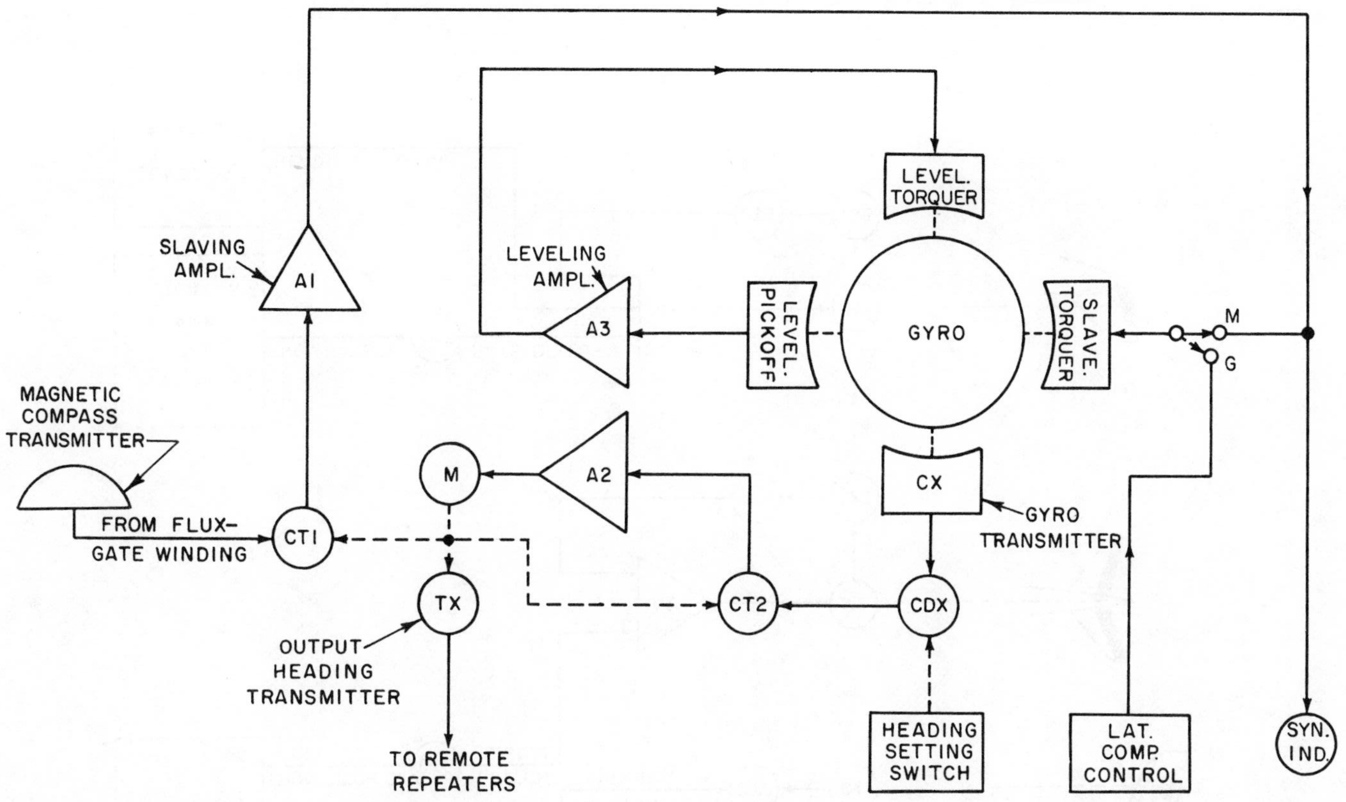

Simplified diagram of an aircraft gyrocompass system.

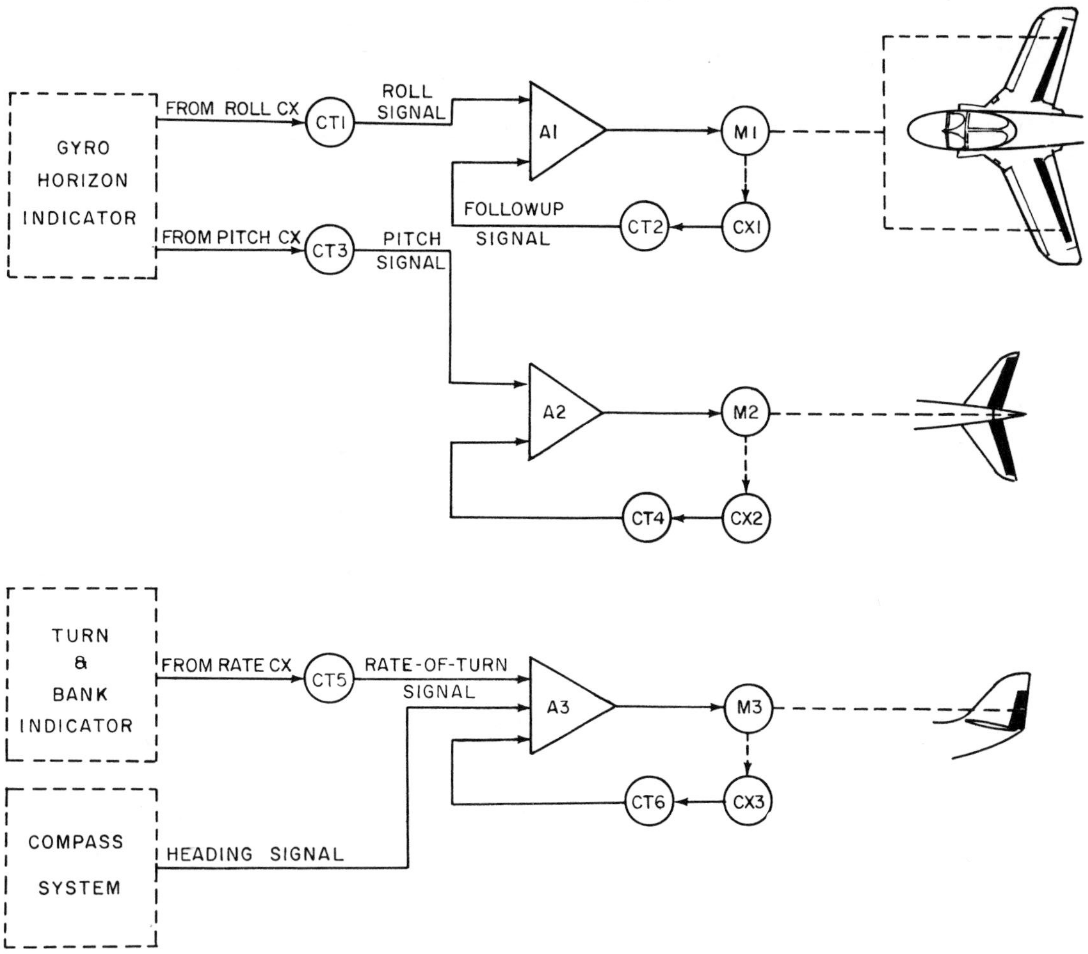

Simplified diagram of an aircraft autopilot system.

Index